HOW

WESTERN FARMERS

ARE

Benefited by Protection.

BY

DAVID H. MASON,

TARIFF EDITOR OF THE CHICAGO INTER-OCEAN.

1875.

HOW WESTERN FARMERS ARE Benefited by Protection.

BY

DAVID H. MASON,

TARIFF EDITOR OF THE CHICAGO INTER-OCEAN.

COMPILED FROM EDITORIAL ARTICLES WHICH HAVE APPEARED IN THE INTER-OCEAN DURING 1874 AND 1875.

He that is first in his own cause seemeth just; but his neighbor cometh and searcheth him. —*Proverbs*, xviii, 17.

Prove all things; hold fast that which is good.—*First Thessalonians*, v. 21.

PUBLISHED BY THE AUTHOR.

CHICAGO:
No. 460 WEST RANDOLPH STREET.
1875.

TO

HENRY C. CAREY,

THE GREATEST OF ALL THE ADVOCATES OF

Tariff-Protection,

THIS LITTLE WORK IS INSCRIBED

BY ITS AUTHOR,

AS A TOKEN OF PROFOUND RESPECT AND AFFECTIONATE REGARD.

TABLE OF CONTENTS.

PREFACE.

My object in writing the articles contained in these pages has been to convince the public generally, and farmers particularly, that the policy of Protection to home industry is promotive of their interests and conducive to their welfare. Our present tariff system has been so wickedly misrepresented, and even willfully lied about, that it is high time the other side should be heard on its merits, and that the people should investigate for themselves.

The principal propositions which I think have been conclusively established in the following pages are these: That farmers have obtained much better prices for their produce, and have been able to export it in larger quantities, under Protection than under partial Free Trade. That the Protective policy operates to bring the manufacturer to the side of the farmer, thus dispensing with useless transportation between the two, and with superfluous middlemen. That the same policy constantly tends, with accelerating force, to enhance the value of land, of labor, and of raw materials, including the produce of the soil, while diminishing the value of manufactured articles—the more, too, as these take on higher forms of reproduction. That the foreign market is secondary in importance, the market at home being the one in which the great mass of domestic products must be sold and consumed. That foreigners either partly or wholly pay a very considerable amount of the duties levied on imports, and thus are forced to contribute to the support of our Government as an offset to the privilege of our markets, diminishing the taxes that otherwise would have to be paid by our own people. That railroad construction and transportation have been cheapened by the tariff on rails. That the "sell dear, buy cheap" maxim is merely a delusive form of words. That an increasing proportion of manufactures enters into our domestic exports under the Protective system. That Protection is highly beneficial to the manufacturer, as well as to the public, notwithstanding that it operates to reduce the prices of his products. And that the price of everything the farmer has to buy has been cheapened by Protection.

For these propositions, and for the facts, figures, and arguments adduced in their behalf, I ask, in the interest of truth and of patriotism, the earnest attention of the reader.

D. H. MASON.

Chicago, November, 1875.

HOW WESTERN FARMERS ARE BENEFITED BY PROTECTION.

CHAPTER I.

OUR MARKETS, THE HOME AND THE FOREIGN.

THE vast importance of the *foreign* market—the dependence of Western farmers upon the *foreign* market—the danger of obstructing the *foreign* market—these are subjects upon which Free Trade writers and speakers lay great emphasis. It never seems to occur to these persons that the *home* market is the vital consideration after all. In the *home* market the bulk of our productions is consumed. The activity of the *home* market forms the basis of individual and national prosperity. Although the foreign market is valuable, that valuableness is not primary, but secondary—valuable as an adjunct or appendix to internal commerce and trade. Foreign countries take only a very small proportion of our products. In the census year 1870 the total amount of our manufactures was $4,232,325,442, and of farm produce $2,447,538,658, or $6,679,864,100 for both. That the immense quantity of articles here represented, except a small fraction, must have been sold and consumed in our home market, is plain from the fact that the aggregate of our exports from the beginning of the Government to June 30, 1875, covering a period of eighty-six years, was just $13,299,706,575, or not quite double the value of our manufactures and farm products in a single year. Take another illustration: According to the agricultural reports of our Government, the four corn crops of 1870, 1871, 1872 and 1873 footed up 4,111,119,000 bushels, and the four wheat crops 997,858,900 bushels. During the same period we exported 84,252,004 bushels of corn, and 1,111,685 barrels of Indian meal, the whole being equal to 89,254,587 bushels, or only 2.17 per cent. of the product, leaving the remainder, or nearly 98 per cent., to find buyers and consumers in the United States. Of wheat, in the same time, we exported 136,516,386 bushels; of flour, 12,193,795 barrels; of bread and biscuit, 46,209,780 pounds: these three forms of the grain representing 198,688,751 bushels, or 19.91 per cent. of the product, so that fully four-fifths of it remained in this country to supply the local demand. Such facts conclusively show that the *home* market, not the *foreign*, is first in order of importance; is the one which offers the most numerous and reliable opportunities for effecting exchanges; and is the one capable ot the largest development. To expand, sustain, and prosper that market are prime objects of the Protective policy.

It is a favorite dogma of the Free Traders that a high tariff erects barriers to commerce, and restricts both imports and exports, thus tending to scarcity. This, however, is nothing but a theory, and is a libel upon the facts, as we shall prove. Below, in contrast, are two tabular statements of exports and imports, compiled from the official reports, covering the last twenty-eight fiscal years:

Fourteen Years under Partial Free Trade.				Fourteen Years under Protection.			
Fiscal Years.	Domestic Exports.	Import Entries.	Exports of Imports.	Fiscal Years.	Domestic Exports.	Import Entries.	Exports of Imports.
1848	$132,904,121	$154,998,928	$21,128,010	1862	$213,069,519	$205,862,518	$16,869,466
1849	132,666,955	147,857,439	13,088,865	1863	305,884,998	252,919,920	26,123,584
1850	136,946,912	178,138,318	14,951,808	1864	320,035,199	329,565,115	20,256,940
1851	196,689,718	216,224,932	21,698,293	1865	306,306,758	234,434,167	30,390,365
1852	192,368,984	208,296,855	17,289,382	1866	550,684,277	437,640,354	14,742,117
1853	213,417,697	267,978,647	17,558,460	1867	438,577,312	417,831,571	20,611,508
1854	253,390,870	301,494,094	23,748,514	1868	454,301,713	371,624,808	22,601,126
1855	246,708,553	261,468,520	28,448,293	1869	413,961,115	437,314,255	25,173,414
1856	310,586,330	314,639,942	16,378,578	1870	499,092,143	462,377,587	30,427,159
1857	338,985,065	360,890,141	23,975,617	1871	562,518 651	541,493,708	28,459,899
1858	293,758,279	282,613,150	30,886,142	1872	549,219,718	640,338,766	22,769,749
1859	335,894,385	338,768,130	20,895,077	1873	649,132,563	663,617,147	28,149,511
1860	373,189,274	362,166,254	26,933,022	1874	693,039,054	595,861,248	23,780,338
1861	227,966,169	352,739,287	20,645,427	1875	643,081,433	553,906,253	22,374,710
Tot.	$3,385,473,312	$3,748,274,637	$297,625,488	Tot.	$6,598,904,453	$6,144,787,417	$332,729,886

These official statistics emphatically contradict the Free Trade dogma. During the Protective period we find a vast expansion of both exports and imports; not the alleged restriction. The total of domestic exports, as compared with that in the period of partial Free Trade, exhibits an increase of 94.9182 per cent., while a like contrast between the two footings of import entries shows an augmentation of 63.9364 per cent. The figures of exports of imports also possess great significance. Although the excess in the last fourteen years over the previous fourteen was $35,104,398, equal to an advance of 11.7948 per cent., the ratio of exports of imports to import entries in the low-tariff period was 6.9403 per cent., but in the high-tariff period only 5.4148 per cent., making it very clear that *under Protection we retained for home consumption a much larger proportion of our imports than we did under partial Free Trade.*

An attempt is sometimes made to parry the force of this strong statistical argument by saying that population in this country has more than doubled since the passage of the tariff of 1846, and that, vast as our exports and imports have been in recent years, they have not kept pace with the growth of numbers, but are relatively to the mass of individuals less now than they were a quarter of a century ago. Apparently this objection is well founded, yet it will not bear the test of contact with the realities of the case. Conclusively to settle the point, we have obtained, through a very tedious and laborious series of trial calculations, the exact per cent. of annual increase of population for the decades ending severally in 1850, in 1860, and in 1870. These rates are so minutely accurate, that, on multiplying the census enumeration by its connected per centum, and then, after pointing off and discarding the decimal fraction, adding the product to the multiplicand for a new multiplicand, and repeating the process for the remaining years of the decennial period, the final result will be found identical with the official count of the people for the current decennary. It being impossible to know what will be the number of inhabitants in 1880, we have assumed that the rate of increase since 1870 has been just what it was during the decade 1850–60—an exceedingly liberal allowance, considering what the superintendent of the ninth census has to say about the probable increment. Having thus obtained a trustworthy statement of the population in each year, we have employed it in computing the *per capita* of domestic exports and of import entries, annually, for the period embraced in our tables above. Here are the results:

Fourteen Years under Partial Free Trade.					Fourteen Years under Protection.				
Years.........	Population in each year...	Per cent. of annual increase	Domestic exports per capita......	Import entries per capita......	Years.........	Population in each year..	Per cent. of annual increase......	Domestic exports per capita......	Import entries per capita. .....
1848	21,812,861	3.112576	$6.09.3	$7.10.6	1862	32,752,651	2.060815	$6.50.5	$6.28.5
1849	22,491,802	....	5.45.4	6.57.5	1863	33,427,622	....	9.15.1	7.56.6
1850	23 191,876	....	5.90.5	7.68.1	1864	34,116,503	....	9.38.1	9.66.0
1851	23,908,654	3.090646	8.22.7	9.12.7	1865	34,819,581	....	8.79.7	6.73.3
1852	24,647,585	....	7.80 5	8.45.1	1866	35,537,148	....	15.49.6	12.31.5
1853	25,409,354	....	8 39.9	10.54.6	1867	36,269,502	....	12.09.2	11.52.0
1854	26,194,667	....	9.67.3	11.12.8	1868	37,016,949	....	12.27.3	10.03.9
1855	27,004,251	....	9.13.9	9.68.3	1869	37,779,800	....	10.95.7	11.57.5
1856	27,838,856	....	11.15.7	11.30.2	1870	38,558,371	....	12.94.4	11.99.2
1857	28,699,256	....	11.81.2	12.57.5	1871	39,750,c73	3.c90646	14.15.1	13.62.2
1858	29,586,248	....	9.92.9	9.55.2	1872	40,978,607	....	13.40.3	15.62.6
1859	30,500,654	....	11.01.3	11.10 7	1873	42,245,110	....	15.36.6	15.70.9
1860	31,443,321	....	11.86.8	11.51.8	1874	43,550,756	....	15.91.3	13.68.2
1861	32,091,309	2.060815	7.10.4	10.99.2	1875	44,896,745	....	14.32.4	12.33.7
Tot.	374,820,694	....	$9.03.2	$10.00.0	Tot.	531,699,418	....	$12.41.1	$11.55.7

There is no escape from the proofs offered by these figures. They show, beyond room for doubt, that domestic exports and import entries increased faster under Protection than they did under partial Free Trade, even distributed *per capita*, the growth of commerce and trade being far more rapid than the growth of population. The average per individual inhabitant, during the second period of fourteen years, was greater by $3.37.9 of domestic exports, equal to 37.4114 per cent. more; and by $1.55.7 of import entries, equal to 15.57 per cent. increase. Are these evidences that the Protective policy restricts either exports or imports, and tends to scarcity? Rather, do they not contradict such a theory with all the conclusive force of positive knowledge? Our mathematical refutation of the absurd dogma of the Free Traders is complete.

Under the series of tariffs since 1861 we have sold more abroad, and bought more abroad, of a wide range of commodities, than we ever did before in a like term of years. The vastness of these exchanges may be made appreciable by a simple illustration. Our domestic exports, from *the beginning of the Government* to June 30, 1861, amounted to $6.700,802,122; for the fourteen years ending June 30, 1875, to $6,598,904,453, the latter value being 98.4793 per cent. of the former, indicating a prodigious growth of *export* power. Further, our imports retained for consumption, during the seventy-two years, amounted to $7,487,067,366; during the last fourteen years, to $5,812,057,531, or to 77.628 per cent. of the former value, exhibiting an immense increase of *import* power. These marvelous facts and figures possess the attributes, not of restriction or obstruction, but of expansion. What an abundance must have remained to be consumed at home, when we could spare so much to be sent abroad! What a remarkable capacity to buy and consume foreign goods is made manifest by the stupendous value of imports retained for our own use!

The explanation of this extraordinary growth of both export and import trade is to be found in the activity, enterprise, and prosperity of the *home market*, through the beneficial operation of the Protective policy upon domestic industry. That policy encouraged manufactures on our own soil; diversified employments; created a demand for labor; advanced the rate of wages, and conferred purchasing power upon the masses of the people. It is the expenditure of earnings that energizes commerce and trade. A population that is steadily employed and well paid can buy and consume; but not otherwise. The sum of social misery among a people can be measured by their inability to obtain wages. Regular employment and labor fully compensated are the fruitful parents of general prosperity, content, and cheerfulness. Where there is work for the hands of men, there is work for their teeth, clothes for their bodies, shelter for their heads, fuel for their warmth, instruction for their minds, and progress for their condition. The Protective policy, by conferring purchasing power, caused an increasing demand for all these things; and, by stimulating the productive forces, satisfied that demand. Services easily found employment, and the circulation of commodities from hand to hand became rapid. The home market, formerly sluggish, was converted into a scene of enterprise and thrift; abundance became the order of the day; and both exports and imports expanded to unprecedented values. *A prosperous foreign trade is inseparably connected with a prosperous home market.* There must be a large development of *internal* commerce before there can be a large development of *external* commerce. No more than an individual can a nation exert great strength outwardly unless such strength exists inwardly. To neglect the home market, by reaching out after the foreign, is to neglect the substance to pursue a phantom.

CHAPTER II.

OUR EXPORTS OF WHEAT.

IT is a dogma of the Free Traders that Western farmers, of all classes, are the greatest sufferers by the Protective policy—that our present tariff system operates to reduce the price of everything the farmer has to sell, and to increase the price of everything he has to buy. Although these allegations are contradicted by experience, they have obtained credence in various quarters through the mere dint of reiteration that was not immediately antagonized and shown to be false. But the people have been too much imposed on any longer to accept statements about facts unless accompanied by the facts themselves. The day has gone by in which unsupported assertions exerted a strong influence upon the popular mind. Only arguments and proofs find favor now.

We intend to make that kind of appeal in this article. Taking wheat as an illustration, we shall show that our farmers sell more of that grain abroad, have a steadier market for it, and receive better prices, under the policy of Protection, than they did under the policy of partial Free Trade. The figures offered below are derived from the Commerce and Navigation Reports, published by our own Government, and, being official, are therefore competent and reliable witnesses. First, we give the total quantity and value of wheat exported in each of the years specified, to which we have added a calculation of the average price per bushel in the port of departure—a calculation the accuracy of which any schoolboy can verify—the fraction of a cent being extended to three decimal places for the sake of greater exactness. The first table covers a period of thirteen years under partial Free Trade; the second, an equal period under our Protective system. Observe the signal contrast which contradicts at every point the allegation we antagonize.

Exports under Partial Free Trade.				Exports under our Protective System.			
Fiscal Years.	Bushels.	Values.	Average price per bush.	Fiscal Years.	Bushels.	Values.	Average price per bush.
1849.........	1,527.534	$1,756,848	$1.15.012	1862........	37,289,572	$42,573,295	$1.14.169
1850.........	608,661	643,745	1.05.764	1863.........	36,160,414	46,754,195	1.29.296
1851.........	1,026,725	1,025,732	.99.903	1864.........	23,681,712	31,432,133	1.32.727
1852........	2,694,540	2,555,209	.94.829	1865.........	9,937,152	19,397,197	1,95.200
1853........	3,890,141	4,354,403	1.11.934	1866.........	5,579,103	7,842,749	1.40.574
1854........	8,036,665	12,420,172	1.54.545	1867.........	6,146,411	7,822,555	1.27.270
1855.........	798,884	1,329,246	1.66.388	1868.........	15,940,899	30,247,632	1.89.748
1856.........	8,154,877	15,115,661	1.85.357	1869.........	17,557,836	24,383,259	1.38.874
1857.........	14,570,331	22,240,857	1.52.645	1870.........	36,584,115	47,171,229	1.28.939
1858.........	8,926,196	9,061,504	1.01.516	1871.........	34,304,906	45,143,424	1.31.595
1859.........	3,002,016	2,849,192	.94.909	1872.........	26,423,080	38,915,060	1.47.277
1860.........	4,155,153	4,076,704	.98.112	1873.........	39,204,285	51,452,254	1.31.241
1861.........	31,238,057	38,313,624	1.22.650	1874.........	71,039,928	101,421,459	1.42.767
Totals.....	88,629,780	$115,742,897		Totals....	359,849,413	$494,556,441	
Ann'l av.	6,817,676	$8,903,300	$1.30.591	Ann'l av.	27,680,724	$38,042,803	$1.37.434

These statistics conclusively show that we exported, in the Free Trade period, only 24⅝ per cent. of the quantity and only 23⅖ per cent. of the value that were exported in the Protective period; or, to state the case differently, we exported more by 271,219,633 bushels and $378,813,544 in the second thirteen years than we did in the first thirteen. Besides, we had gained in average price 6.843 cents per bushel. More than that, the quantity and value of wheat exported in the years 1862–74 are much greater than the quantity and value exported from the beginning of the Government to July 1, 1861, covering a period of *seventy-two years*. Even in the three fiscal years, 1862, 1863, and 1864, we exported 8,501,918 bushels more than during the *entire period of thirteen years* under partial Free Trade. Taking 1874 by itself, we exported in that twelvemonth 80⅙ per cent. of the quantity, and 87⅝ per cent. of the value, exported between June 30, 1848, and July 1, 1861, at an average increase of 12⅙ cents per bushel. It is very evident from these facts that the policy of Protection to home industry does not act repressively upon the farmer's export markets, as regards either quantity or price. Partial Free Trade certainly would have done so, because that policy, by dwarfing the number and extent of our manufacturing establishments, would have diminished the number of persons in this country who were consumers without being producers of wheat, and thus would have left a larger surplus for export, which, as a more copious supply in the presence of a demand not augmented, must, through increased competition for sale abroad, have resulted in a reduction of prices. Those low prices, unremunerative to the farmer, would have reacted injuriously upon agriculture ; and he would soon have ceased to grow wheat beyond what he could sell at a profit. Thus does Free Trade dry up his resources, while his expenses are greatly increased ; since all manufactured articles—the necessaries as well as the conveniences of life—rise in cost as producer recedes to a greater distance from consumer.

Under Protection, the farmer finds a readier and steadier market for his surplus produce. Observe how our statistical tables support this statement. Notice the hop-skip-and-jump sort of demand abroad exhibited during the first thirteen years, and then see the copious and expanding exports in the second period, interrupted, it is true, for a while, by the war of secession, but afterward mounting up by strides. While hostilities were raging, soldiers, engaged by the hundred thousand in the work of destruction, had to be fed; and when they returned to the arts of peace, a half-starved South, with desolated farms and almost exhausted supplies of food, had to be provided. These disturbances over, and commerce once more returned to its wonted channels, exports of wheat augmented with strengthening rapidity. On the contrary, the period of partial Free Trade displays very great irregularity and unsteadiness. In 1853 the export was 3,890,141 bushels; in 1854 the quantity bounded to 8,036,665 ; in 1855 it dropped like a shot to 798,884; in 1856 rushed up to 8,154,877; in 1857 was lifted, freshet-like, to 14,570,331 ; in 1858 tumbled to 8,926,196 ; and in 1859 further to 3,002,016. We look in vain for such marked, zig-zag fluctuations under our system of Protection.

But our statistics, as presented, exhibit only part of the facts. All exports from the Pacific ports stand for gold values. Therefore, we must separate and subtract those from the currency values in order to obtain an accurate statement of the latter. We have done so. The following table gives a detailed exhibit of the exports of wheat, at currency values, from all Atlantic and Gulf ports, showing number of bushels exported, percentage of total exports of United States, their values, percentage of total value exported, and average price per bushel :

Fiscal Years.	Bushels.	Per centage of total quantity exported.	Values.	Per centage of total value exported.	Average price per bushel.
1862......	35,749,502	95.870	$40,876,683	96.015	$1.14.342
1863......	34,383,147	95.085	44,796,485	95.813	1.30.286
1864......	21,887,946	92.426	29,724,187	94.567	1.35.792
1865......	9,881,343	99.438	19,284,430	99.419	1.95.160
1866.....	*...........	*.........	*.............	*.........	*.........
1867	1,254,789	20.415	2,419,553	30.930	1.92.825
1868.....	10,388,449	65.168	20,277,002	67.037	1.95.188
1869......	11,968,089	68.164	18,092,651	74.205	1.51.174
1870......	28,619,293	78.229	38,926,139	82.521	1.36.014
1871.....	28,038,788	81.734	37,637,325	83.373	1.34.233
1872......	23,803,990	90.088	35,276.985	90.651	1.48.198
1873......	22,325,152	56.946	33,230,745	64.586	1.48.898
1874....	57,843,253	81.424	84,602,636	83.417	1.46.262
Totals.	286,143,741	81.068	$405,144,821	83.507	$1.41.588

A comparison between these prices and those in the period of partial Free Trade indicates a large balance of advantages in favor of the policy of Protection. The average export price is some eleven cents higher per bushel. We know that the common reply is that these represent prices in depreciated paper money, while those under the tariffs of 1846 and 1857 represent gold; but there is no force in this objection. Our farmers who sold their wheat in the Chicago market during the years before the war must have a vivid recollection of the "red-dog," the "stump-tail," and the "wild-cat" currency of the period. It is sheer nonsense to talk of such paper money having been equivalent to gold; a circulating medium whose value was fixed from time to time, by edicts of the railroad companies, as receivable by them for freights and fares; a currency that often was uncurrent a hundred miles from home, and that would not buy New York exchange without suffering a considerable shave. Who wants to give up greenbacks or national bank notes in order to get back that miserable stuff in which our farmers were paid for their wheat?

There is another very important phase to the subject. During the war, when the gold premium was so high, our farmers, who had got deeply in debt amid the great prosperity claimed to have been conferred upon them by partial Free Trade, were rapidly clearing off mortgages and other incumbrances upon their property. Greenbacks being legal tenders, every dollar of them devoted by our farmers to the payment of debts contracted before the war was equivalent to a dollar in coin, no matter how extravagant the

* Not separately stated in the Commerce and Navigation Report for that year.

premium on gold might be. Our farmers obtained the full benefit of the high prices at which they sold their produce.

Besides, the farmer to-day is able to pocket a larger proportion of the current price for wheat than he did in the years before the war. Then a very considerable portion of the grain sent to Chicago arrived by wagon or by canal. The cost of getting to market ate a large hole in the farmer's profits, and sometimes left none at all. Now the facilities for transportation are much more numerous, much more extensive, and much cheaper. This lessened cost of conveyance adds value to the price of wheat at the farm, and such increase is realized by the producer. Moreover, this solid benefit is an outgrowth of the policy of Protection, which gave, as part of the general stimulus to domestic industry, a powerful impulse to the construction of railroads, accompanied by large improvements in rolling stock, road-beds, bridges, station-buildings, and accessories for the cheap handling of grain.

Taking wheat as an example, we think we have conclusively shown that Western farmers are not repressively or injuriously acted upon by the policy of Protection, but have received from it some very important and substantial benefits. Then why should not farmers favor Protection to home industry?

CHAPTER III.

EXPORTS OF WHEAT FLOUR.

ON turning to the exports of wheat flour, we find a state of facts which again and further contradicts the dogmatic assertions made by the Free Traders about the depressing influences of Protective tariffs upon agricultural prosperity. Below are the official figures of quantities and values of flour exported from the United States in each of twenty-six consecutive years, to which we have added a calculation of the average price per barrel, the fraction of a cent being extended to three decimal places for the sake of greater exactness.

Thirteen Years under Partial Free Trade.				Thirteen Years under our Protective System.			
Fiscal Years.	Barrels.	Values.	Average price per barrel.	Fiscal Years.	Barrels.	Values.	Average price per barrel.
1849.........	2,108,013	$11,280,582	$5.35.129	1862.........	4,882,033	$27,534,295	$5.63.990
1850.........	1,385,448	7,098,570	5.12.366	1863.........	4,390,055	28,366,069	6.46.144
1851.........	2,202,335	10,524,331	4.77.871	1864.........	3,557,347	25,588,249	7.19.310
1852.........	2,799,339	11,869,143	4 23.998	1865.........	2,604,542	27,222,031	10.45.175
1853.........	2,920,918	14,783,394	5.06.122	1866.........	2,183,050	18,396,686	8.42.706
1854.........	4,022,386	27,701,444	6.88.682	1867.........	1,300,106	12,803,775	9.84.825
1855.........	1,204,540	10,896,908	9.04.653	1868.........	2,076,423	20,887,798	10.05.950
1856.........	3,510,626	29,275,148	8.33.901	1869.........	2.431,873	18,813,865	7.73.637
1857.........	3,712,053	25,882,316	6.97.253	1870.........	3,463,333	21,169,593	6.11.249
1858.........	3,512,169	19,328,884	5.50.340	1871.........	3,653,841	24,093,184	6.59.393
1859.........	2,431,824	14,433,591	5.93.529	1872.........	2,514,535	17,955,684	7.14.076
1860.........	2,611,596	15,448,507	5.91,535	1873.........	2,562,086	19,381,664	7.56.480
1861	4,323,756	24,645,849	5.70.010	1874.........	4,094,094	29,258,094	7.14.641
Totals....	36,745,003	$223,168,667		Totals....	39,713,318	$291,470,987	
Ann'l av.	2,826,539	$17,166,821	$6.07.344	Ann'l av.	3,054,871	$22,420,845	$7.33.938

Here we see that the quantity was larger, the value was greater, and the price was higher, under the policy of Protection than under the policy of partial Free Trade. Are these evidences that the interests of agriculture are injuriously affected by our present tariff system? The average price during the second thirteen years was $1.26 594 more per barrel than during the first thirteen.

Does this show that the Protective policy is detrimental to the prosperity of Western farmers? If the 39,713,318 barrels of flour exported in the Protective period had been sent abroad at the same average price received for the 36,745,003 barrels exported in the period of partial Free Trade, then the total export value would have amounted to only $241,196,454, instead of $291,470,987—the sum actually realized—making a difference, which would have been a loss, of $50,274,533. Is that value of *more than fifty million dollars*, which represents a positive gain, to be regarded as part of the baneful effects produced upon our agriculture by the series of tariffs since 1861? Are better prices and larger sales usually looked upon as very serious evils, and as oppressive to the grain-grower? We put these questions to the common sense of the reader.

The case will appear in a light still stronger and clearer if we subtract the exports of flour from the Pacific ports, at gold prices, from the exports from the Atlantic and Gulf ports, at currency prices, and thus obtain an accurate statement of the latter. The following table gives a detailed exhibit of such exports.

Fiscal Years.	Barrels.	Percentage of total quantity exported.	Values.	Percentage of total value exported.	Average price per barrel.
1862......	4,778,311	97.875	$26,920,234	97.768	$5.63.384
1863......	4,238,219	96.541	27,509,183	96.979	6.49.074
1864......	3,386,192	95.189	24,677,190	96.440	7.28.759
1865......	2,548,012	97.830	26,707,171	98.109	10.48.155
1866.....	*...........	*.........	*.............	*.........	*............
1867......	1,004,110	77.233	11,171,241	87.250	11.12.552
1868......	1,781,769	85.810	19,126,196	91.566	10.73.439
1869......	2,016,320	82.912	16,604,069	88.254	8.23.484
1870......	3,102,192	89.572	19,425,761	91.763	6.26.195
1871......	3,429,697	93.866	22,766,553	94.494	6 63.807
1872......	2,224,751	88.476	16,184,006	90.133	7.27.454
1873......	2,259,840	88.203	17,861,622	92.157	7.90.393
1874......	3,389,119	82.781	24,870,819	85.005	7.33.843
Totals.	34,158,532	91.459	$253,824,045	93.395	$7.43.077

These are the quantities, values, and prices in which Western farmers are directly interested. The proof is that the average export price under the policy of Protection has been $1.35.733 more per barrel than under the policy of partial Free Trade. *How terribly our farmers have suffered from our present tariff system!* An increase of 22⅓ per cent. in the average export price of flour

* Not separately stated in the Commerce and Navigation Report for that year.

is, from the farmer's standpoint, a lamentable evidence of tariff spoliation—*according to the Chicago Tribune!*

The ready answer comes back, that the above prices stand for depreciated and irredeemable paper-money; are reckoned in that worthless trash called greenbacks; while the prices under partial Free Trade are really higher, although nominally lower, because they represent the purchasing power of gold. It is about time to sweep this pernicious error into the general pile of useless rubbish. Here is what was thought and written about this wonderful gold-base currency, when it was in circulation. We copy from the "Sixth Annual Review of the Commerce, Manufactures, and the Public and Private Improvements of Chicago, for the year 1857," published from the columns of the Chicago *Daily Press.* Our extract appears on page 25, as follows:

> Till the 12th of last September, for nearly four years past, the range for exchange on New York has been three-quarters of one per cent., the variation being from 1½ @ ¾ per cent. premium. As most of our readers have had occasion to feel, since September the price has been ruinously high; but such has been the determination of our mercantile community to maintain their honor untarnished that enormous sacrifices have been cheerfully made to meet maturing obligations. *Ten per cent. has been the highest price charged by the regular bankers; but for many days during the last three months they have had none to sell even at these figures. Twelve and even fifteen per cent. has been paid by parties to the brokers for small amounts, rather than have notes protested.*
>
> These enormous prices for exchange have cost our city an immense amount of money; but as a whole there can be no doubt that it has been a great and positive advantage to our trade to do it. IT WAS CAUSED BY THE IMPOSSIBILITY OF CONVERTING THE BILLS OF THE ILLINOIS AND WISCONSIN BANKS INTO COIN. * * * For a time no one dared to keep an account in St. Louis, for nearly all the banks and bankers failed or were considered in a most precarious condition.

Such was the paper money, *equivalent to gold,* in which our farmers were paid for their grain, and which, as is asserted, possessed *a greater purchasing power than greenbacks.* What a mockery of the facts!

Even if we take the false method usually employed to reduce currency prices to equivalent gold, we shall arrive at results which emphatically contradict the assertion that the assumed coin values, embodied in average export prices under partial Free Trade, represent higher coin values than the average export prices in currency under the Protective policy. For example, take the following:

COMPARATIVE STATEMENT OF AVERAGE EXPORT PRICES PER BARREL.

Fiscal Years.	Gold Prices.	Fiscal Years.	PacificCoast Gold Prices.	Currency Prices.	Gold value of a paper dollar.	Currency prices reduced to gold.
1858	$5.50.340	1871	$5.91.865	$6.63.807	88.7	$5.88.796
1859	5.93.529	1872	6.11.379	7.27.454	89.4	6.50.343
1860	5.91.535	1873	5.02.915	7.90.393	87.3	6.90.013
1861	5.70.010	1874	6.22.331	7.33.843	89.3	6.55.321

These statistics are given merely for the purpose of fighting the Free Traders on their own ground. We place no reliance in such reductions of currency to gold, *because the gold value of a paper dollar, averaged for a year, for a month, or even for a single day, can be obtained only by violating arithmetical rule.* Of course, the outcome of such a process must be false. Still, as Free Traders insist on thus violating arithmetical rule, we are justified in showing that even there do they find no support for their argument. Let it be observed, therefore, that the currency prices reduced to gold are higher than those under partial Free Trade; and that, with the exception of the year 1858, they are higher than the gold prices of the Pacific Coast. Now, in view of all these exhibits, by what right, or with what reason, can it be asserted that Western farmers are repressively and injuriously acted upon by the policy of Protection to home industry?

CHAPTER IV.

EXPORTS OF INDIAN CORN.

IN continuance of our plan of giving *facts* instead of *assertions about the facts*, as Free Trade journals do, we will next show that farmers have received, under the policy of Protection to home industry, higher prices for corn, and that larger quantities of it have been exported, than under the policy of partial Free Trade. Here are the official figures of our domestic exports of Indian corn for the past twenty-six years, to which we have added a calculation of the average price per bushel.

Thirteen Years under Partial Free Trade.				Thirteen Years under the Policy of Protection.			
Fiscal Years.	Bushels.	Values.	Average price per bushel.	Fiscal Years.	Bushels.	Values.	Average price per bushel.
1849.....	13,257,309	$7,966,369	$0.60.090	1862......	18,904,898	$10,387,383	$0.54.945
1850.....	6,595,092	3,892,193	.59.016	1863......	16,119,476	10,592,704	.65.714
1851......	3,426,811	1,762,549	.51.434	1864......	4,096,684	3,353,280	.81.853
1852......	2,627,075	1,540,225	.58.629	1865......	2,812,726	3,679,133	1.30.803
1853......	2,274,909	1,374,077	.60.401	1866......	13,516,651	11,070,395	.81.902
1854......	7,768,816	6,074,277	.78.188	1867......	14,889,823	14,871,092	.99.874
1855......	7,807,585	6,961,571	.89.164	1868	11,147,490	13,094,036	1.17.462
1856......	10,292,280	7,622,565	.74.061	1869......	7,047,197	6,820,719	.96.786
1857......	7,505,318	5,184,666	.69.080	1870......	1,392,115	1,287,575	.92.490
1858......	4,766,145	3,259,039	.68.379	1871......	9,826,309	7,458,997	.75.908
1859......	1,719,998	1,323,103	.76.925	1872......	34,491,650	23,984,365	.69.537
1860.....	3,314,155	2,399,808	.72.411	1873......	38,541,930	23,794,694	.61.737
1861......	10,678,244	6,890,865	.64.532	1874......	34,434,606	24,769,951	.71.933
Totals.	82,033,737	$56,251,307		Totals.	207,221,555	$155,164,324	
An. av.	6,310,287	$4,327,024	$0.68.571	An. av.	15,940,120	$11,935,717	$0.74.878

These statistics establish the fact that, during thirteen years of Protection, the average export price of corn was very nearly 6⅓ cents per bushel more than it was during the previous thirteen years of Free Trade policy. Is that evidence that Western farmers have been plundered by the series of tariffs since 1861? If the 207,221,555 bushels exported in the period 1862–74 had obtained no greater export price, on an average, than $0.68.571, such having been the average for the period 1849-61, then the amount realized would have been only $142,093,891.10, instead of the $155,164,324 actually received, so that the loss would have been $13,070,432.90. Does that increase of *more than thirteen million dollars* constitute part of the proofs that the country is going from bad to worse, and that the people generally, and the Western farmers particularly, are being impoverished, all on account of Protective duties? The exports in the second period, at the higher average price, were 152.6 per cent. larger than the exports in the first period, at the lower price. Is that copious and extraordinary gain in export quantity, coupled with a more remunerative value, further indication of the baneful effects produced upon Western farmers by our present tariff system? Rather, may it not be that the highly-wrought picture of wrongs and outrages suffered by the agricultural classes, under the policy of Protection to home industry, as paraded before the public, with a great blare of announcing trumpets, by Free Traders, is merely a picture painted by the imagination—is only a delineation of non-existent circumstances which are at sixes and sevens with the truth? Between the *real* facts and the *alleged* facts there is an utter want of correspondence. So soon as the children of experience are put on the witness-stand, they testify against the Free Traders. It is sheer nonsense to argue the tariff question, as some do, in such a way as virtually to maintain that assertions are stronger supports of argument than facts.

The swift answer comes back that, while farmers may have obtained higher prices for their produce under the policy of Protection than they did under the policy of partial Free Trade, yet "the American producer has received in exchange for his exports from one-quarter to one-third less in quantity in other commodities, such as iron, and cotton and woolen clothing, than he did between 1846 and 1861." This mode of putting the case is the same old dogmatic plan—assertion without proof. In reply, we quote from the editorial columns of the *Industrial Age*, of Chicago, May 1, 1875, as follows:

It is surprising to see a blanket monopoly sheet, that pretends to wield influence, boldly assert that the prices of goods are one-third higher now than in 1857, when a comparison of prices then with the commercial columns of this same paper flatly contradicts the statement. The editor must think his readers are all fools, who have no means of finding out what goods sold for in 1857. Potter Palmer sold calicoes at one shilling a yard in 1857, and like goods are now selling at eight cents. We can name hundreds of articles, the prices of which are from twenty to fifty per cent. less in currency than they were in gold in 1857.

The writer of the above paragraph tells us that he had charge of the prints department in Potter Palmer's establishment some years before the war, and that he has drawn his figures from personal recollection of daily contact with the facts. It thus appears that the Free Traders, in their assertions about comparative prices, have put the cart before the horse. *The higher prices of general manufactures existed under partial Free Trade, and do not now under the policy of Protection.*

It is very clear to us that Western farmers received the full benefit of the very high prices for agricultural produce during and just after the war. During the fiscal year 1865, when the average export price of wheat was $1.95.160 per bushel; of wheat flour was $10.48.155 per barrel; and of corn was $1.30.803 per bushel, not one of these prices represented a delusive and unsatisfactory value, but one that was substantial and profitable. When the price of gold was $2.85, and a paper dollar was worth very little more than thirty-six cents in coin, it was worth 100 cents in coin to the farmer when applied to the payment of an old debt.

What has contemptuously been called the depreciation of our currency, so far from being an injury to the agricultural classes, has been to them a positive blessing. When the premium on gold is high, they receive higher currency prices for their breadstuffs; and the currency prices of other commodities being cheaper relatively than the currency prices of grain, the money received by farmers for the latter will exchange for larger quantities of those other commodities than when greenbacks approach par with coin. It is folly to say that farmers are victimized by such results as these; for it is folly to assert that *all* prices are governed by the fluctuations of the gold market. When Mr. McCulloch was Comptroller of the Currency, he said, in his report dated Nov. 25, 1864: "When gold sold in Wall street, on the first of July last, at 185 premium, many of the best stocks, as well as productive real estate, were no higher than they have been upon a coin basis." This single proof is sufficient to establish the fact that general prices do not follow the price of gold; and it is in full concurrence with our position that when a high gold premium makes a high export price in currency for grain, the farmer's higher currency price will purchase a larger quantity of other commodities, the prices of which are not affected in like manner.

CHAPTER V.

EXPORTS OF INDIAN MEAL.

OUR exhibit of the exports of corn would not be complete without a like statement of the exports of meal. In neither of these forms of the breadstuff has the quantity exported from the Pacific Coast at gold values been sufficiently large to make any sensible impression upon average prices; hence it is hardly worth while to give those exports in detail. Omitting the fiscal year 1866, for which there are no official statistics of exports by customs-districts, the corn exported from the United States, on the Pacific side of the continent, during the period between June 30, 1861, and July 1, 1874, amounted to only 7,552 bushels, valued at $7,037, equato an average of $0.93.181 per bushel; and of Indian meal, only 2,571 barrels, invoiced at $17,539, equivalent to an average of $6.82.186 per barrel. These figures furnish evidence that the coin prices of commodities on the Pacific Slope constitute no proper standard by which to measure the coin value of the currency prices in other parts of the country; for the above gold prices of corn and meal are much higher than the greenback prices elsewhere, showing they result from a combination of circumstances local in influence and distinct in character.

Below, copied from the Commerce and Navigation Reports of the United States, are the total quantity and value of Indian meal exported in each of the last twenty-six years, together with the average price per barrel.

THIRTEEN YEARS UNDER PARTIAL FREE TRADE.				THIRTEEN YEARS UNDER THE POLICY OF PROTECTION.			
Fiscal Years.	Barrels.	Values.	Average price per barrel.	Fiscal Years.	Barrels.	Values.	Average price per barrel.
1849......	405,169	$1,169,625	$2.88.676	1862......	253,570	$778,344	$3.06.954
1850	259,442	760,611	2.93.172	1863......	257,948	1,013,272	3.92.820
1851......	203,622	622,866	3.05.893	1864......	262,357	1,349,765	5.14.476
1852.....	181,105	574,380	3.17.153	1865......	199,419	1,489,886	7.47.113
1853......	212,118	709,974	3.34.660	1866......	237,275	1,129,484	4.76.023
1854......	257,403	1,002,976	3.89.652	1867......	284,281	1,555,585	5.47.200
1855......	267,208	1,237,122	4.62.981	1868......	336,508	2,068,430	6.14.675
1856......	293,607	1,175,688	4.00.429	1869.....	309,867	1,656,273	5.34.511
1857......	267,504	957,791	3.58.421	1870......	187,093	935,676	5.00.113
1858......	237,637	877 692	3.69.341	1871......	212,641	951,830	4.47.618
1859......	258,885	994,269	3.84.058	1872......	308,840	1,214,999	3.93.407
1860......	233,709	912,075	3.90.261	1873......	403,111	1,474,827	3.65.861
1861......	203,313	692,003	3.40.364	1874......	387,807	1,529,399	3.94.371
Totals.	3,280,722	$11,687,072		Totals.	3,640,717	$17,147,770	
An. av.	252,363	$899,006	$3.56.235	An. av.	280,055	$1,319,059	$4.71.000

Here we find the same general result as in the cases of wheat, of wheat flour, and of corn—larger export quantities and higher average prices under the policy of Protection to home industry. Are *these* evidences of the spoliation practiced upon Western farmers through the baneful operation of our present tariff system, as is asserted by Free Traders? If so, what a pernicious influence must be exerted upon agricultural prosperity by increased prices for breadstuffs! The official statistics show, by contrasting two consecutive periods of thirteen years each, that the export price of Indian meal, on an average, was $1.14.765 higher per barrel, or 32⅕ per cent. more under the policy of Protection, than it was under the policy of partial Free Trade. *What a calamity that was to Western farmers! Just think of it! They were plundered by having more money put into their pockets.* That is very sad, indeed! This great wrong can be stopped, however, by returning to the blessings of partial Free Trade.

The average export price for the first period of thirteen years was $3.56.235 per barrel. If the quantity exported in the second period had obtained no higher rate, then the sum realized would have been only $12,969,508.20, or $4,178,261.80 less than what was actually received. How did that gain of more than four million dollars operate to the injury of farmers? Does this large increase of value indicate the alleged scheme and process of spoliation in the series of tariffs since 1861? Can the agricultural classes be robbed by a system of custom-house legislation under which they prosper—under which they get better prices for the produce they sell, and pay lower prices for the finished products they buy?

A multitude of facts, of daily occurrence, show that the tendency of Protective tariffs is to cheapen the prices of manufactured articles. Such a fact is found in the following statement, which we copy from a Philadelphia paper of recent date:

> Our enterprising Philadelphia saw and tool manufacturers, Messrs. Henry Disston & Sons, have a full-page advertisement in the London *Ironmonger*, stating the location of their works and offices, and the nature of their business, and adding that they have an agency at Durham, England, for the sale of their saws, tools, and sheet steel.

Here we see that *Protective* America is beginning to find a market for some of her manufactures on the soil of *Free Trade* England. That could not be, if the necessary effect of Protective duties is to increase cost in production and to enhance prices. Some months ago, the Sheffield (England) *Telegraph* stated that an agent of Messrs. Henry Disston & Sons had arrived in that town with samples of the firm's wares, and was offering to fill an order for any article in the lot at 15 per cent. less than the current prices charged by Sheffield manufacturers—an announcement which created no little stir and uneasiness. In the face of facts like these, we are told by Free Traders that a reduction of the tariff would enable the Sheffield manufacturers to undersell Messrs. Henry Disston & Sons in the United States, and thus cheapen to farmers the prices of saws and tools. Meanwhile, the American firm is seeking customers in the home market of those very rivals who, it is claimed, would overmaster competition in this country were it not for our Protective tariff. Now, we wish Free Traders to answer this question: How is it possible for the Sheffield producers to outdo the Philadelphia producers in cheapening prices in the American market, when those Philadelphia producers are able to outdo those Sheffield producers in cheapening prices in the English market?

CHAPTER VI.

OUR EXPORTS OF POTATOES.

THE fact that farmers obtain better prices for their produce under the system of Protection to home industry than they do under the system of partial Free Trade finds additional confirmation at every succeeding step of inquiry into the subject. Whether we take the cereals, which are cultivated so extensively, and produced in such vast quantities, or resort to the minor crops, for examples, this important and instructive fact is equally established by the investigation. Let us illustrate the case with potatoes. Here are the exports for the last twenty-six years, with the average price per bushel in each year.

Thirteen Years under Partial Free Trade.				Thirteen Years under Protection.			
Fiscal Years.	Bushels exported.	Values.	Average price per bushel.	Fiscal Years.	Bushels exported.	Values.	Average price per bushel.
1849......	109,665	$83,313	$0.75.970	1862......	417,138	$300,599	$0.72.062
1850	155,595	99,333	.63.841	1863......	517,530	413,581	.79.914
1851......	106,342	79,314	.74.584	1864......	463,212	473,911	1.02.310
1852......	148,916	115,121	.77.306	1865......	510,344	724,593	1.41.981
1853......	225,905	152,569	.67.537	1866......	470,753	535,446	1.13.742
1854......	140,575	121,680	.86.559	1867......	512,380	505,875	.98.730
1855*.....	225,013	203,416	.90.402	1868......	378,605	483,395	1.27.678
1856*.....	226,908	153,061	.67.437	1869......	508,249	451,435	.88.822
1857*.....	238,722	205,616	.86.132	1870......	596,968	412,488	.69.097
1858......	242,231	205.791	.84 956	1871.....	553,070	432,815	.78.257
1859......	376,056	284,111	.75.550	1872......	621,537	482,648	.77.654
1860......	380,372	284,673	.74.341	1873......	515,306	498,291	.96.698
1861......	413,091	285,508	.69.115	1874......	497,413	471,332	.94.757
Totals.	2,989,391	$2,273,506	$0.76.053	Totals.	6,562,505	$6,186,409	$0.94.269

* The exports in these years are stated in barrels in the official reports as follows: 81,823 barrels in 1855, at an average price of $2.48.605 per barrel; 82,512 barrels in 1856, at an average price of $1.85.502 per barrel; and 86,808 barrels in 1857, at an average price of $2.36.863 per barrel. For the sake of uniformity, we have reduced barrels to bushels, at the assumed rate of eleven pecks to the barrel.

Here we have, for the Protective period, an increase of 119½ per cent. in aggregate quantity, of 136 per cent. in aggregate value, and of 24 per cent. in average price per bushel of total, showing a very considerable gain in every respect over the period of partial Free Trade. What benefit could be derived by farmers from going back to the previous condition of affairs, under which their rewards were smaller and their sales less? If, as is asserted by the opponents of our Protective policy, the agricultural classes have been shamefully victimized and plundered by the series of tariffs since 1861, is it not exceedingly strange that these evils have been accomplished through adding to the market value of farm produce, by which larger pecuniary returns have been realized from tilling the soil? If Free Traders are correct in their assertion, then, according to the above figures, we have the wonderful paradox of farmers literally robbed into prosperity. Paradoxical robbery of that sort is much more apt to be accepted as a blessing than as a curse. It is very difficult to impress people with the idea that they are oppressed or despoiled by laws under which they have full pockets.

Had the 6,562,505 bushels of potatoes exported during the thirteen years of Protection obtained no higher average price than the $0.76.053 received for the quantity exported during the thirteen years of partial Free Trade, then the total amount realized would have been only $4,990,981.93, or $1,195,427.07 less than the sum actually secured. Now, how is this very large gain in value to be reconciled with the charge that Protective Tariffs operate in such a way as to plunder the farmer of his hard earnings? According to the rules of common sense, a man regards as highly beneficial and entirely satisfactory those legislative influences under which his pecuniary accumulations are enlarged; but, according to the standard of judgment set up by the Free Traders, such influences are to be considered as injurious, and to be stigmatized as a scheme of spoliation. As correctly and cogently might it be said that robust health is a symptom of fatal disease.

Another factor which enters into the export problem needs to be noticed. All exports from the Pacific Coast are stated in gold. By deducting these from the total exports in each of the years in the Protective period, we obtain an exhibit of exports at currency values, as follows:

EXPORTS OF POTATOES AT CURRENCY VALUES.

Fiscal Years.	Bushels exported.	Values.	Average price per bushel.
1862	404,631	$284,217	$0.70.241
1863	490,166	389,922	.79.549
1864	426,695	450,319	1.05.537
1865	486,845	699,992	1.43.781
1866	*	*	
1867	472,967	481,432	1.01.790
1868	347,231	457,235	1.31.680
1869	467,606	426,076	.91.119
1870	542,532	374,391	.69.008
1871	509,913	395 115	.77.487
1872	540,743	438,424	.81.078
1873	445,566	443,950	.99.637
1874	424,504	421,849	.99.375
Totals	5,559,399	$5,262,922	$0.94.667

These are the figures in which Western farmers are specially interested; for their produce is sold for paper money; and, if there is any evidence in comparative statistics, gathered and recorded by Government itself, then there can be no other conclusion than that agricultural industry pays better and thrives more under the policy of Protection than it does under the policy of partial Free Trade. Our present tariff system, instead of plundering the farmer, as is alleged, really fills his pockets.

* Not separately stated in the official report.

CHAPTER VII.

THE FARMERS' WOOL AND IRON QUESTIONS.

THE great scarecrow which Free Traders have set up in the agricultural districts of the West, to drive farmers away from the policy of Protective tariffs, is the assertion that such policy forces them to give more of their produce in exchange for less of manufactured articles than they did under partial Free Trade. They are told, with oracular dogmatism, that every ounce of iron or steel that is contained in their plows, hoes, axes, trace-chains, nails, harvesters, reapers, and other implements is heavily taxed by high import duties, merely in order to feed fat the insatiate greed of a parcel of manufacturing monopolists; and that every tiller of the soil is their dupe and victim. Farmers are also told that every yard of woolen cloth brought from abroad is burdened with weighty entry charges, which largely enhance the cost to consumers, the advance in price being added to the domestic fabrics, as well as to the foreign, so that wool is reduced in purchasing power, with serious loss and damage to the wool-grower. We shall prove that there is not an atom of truth in these allegations.

Let us see how many pounds of wool have been required, under partial Free Trade and under Protection, to pay for a ton of common English bar iron in this country. This contrast will supply an infallible test for all the points in controversy. As we have no quotations at Chicago for iron during the years before the war, we will resort to the great market of New York City. The Finance Reports of the United States for the years 1863, 1873, and 1874 contain monthly quotations of staple articles in that mart of trade for fifty consecutive years, including the two commodities we have chosen to illustrate our position. We compare the six Free Trade years, 1853–58, with the six Protective years, 1869–74, pitting

American wool against common English bar iron. These two periods are peculiarly apt for contrast. They each embody three years just before a panic, a year with a panic, and a year after the panic. Moreover, each term embraces a nearly corresponding number of months during which, from different causes, the price of iron was exceptionally high. In the Free Trade period, the cause was inexorable extortion on the part of foreign manufacturers, who, after having taken advantage of the low tariff of 1846 to prostrate and bankrupt our iron industries, by the means of cheaper and still cheaper prices, until competition had been destroyed, then took a monopoly control of our markets, and advanced their rates to exorbitant figures on immense importations, thus reimbursing themselves for their losses, besides pocketing large profits. This was the process: English bar iron sold at $75@80 per ton in New York, in 1846, under a duty of $25 per ton; but, when that duty was reduced to 30 per cent. *ad valorem*, the price gradually fell to $33.50@41, in 1851, from which point it rose to $62.50@77.50, in 1854, with an import value in that year, retained for home consumption, vastly larger than it had been in 1846 or in 1851. On the other hand, the cause of the very high prices for English iron in 1872 and 1873 was the coal troubles in Great Britain, doubling the price, coupled with extensively successful strikes by laborers for higher wages, thus greatly increasing cost of production. The result appears very decidedly in the declared values of our imports, which are the real market values in the foreign port of shipment. In the fiscal year 1870 we imported from England 101,642,373 pounds of bar iron, at an invoice value of $1,808,825, equal to 1.77.96 cents per pound, or $39.86.2981 per ton. In the fiscal year 1872 we imported 149,-503,607 pounds, at an invoice value of $3,166,636, equal to 2.11.81 cents per pound, or $47.44.5442 per ton. In the fiscal year 1873 we imported 92,796,789 pounds, at an invoice value of $2,867,850, equal to 3.09.05 cents per pound, or $69.22.6361 per ton. Here we find an increase in 1872 of 19.02 per cent. in invoice value over what it was in 1870; of 45.91 per cent. in 1873 over what it was in 1872; and of 73.66 per cent. in 1873 over what it was in 1870. Such were the advances in price made by the manufacturers to their regular customers in the foreign market of production, before a cent could be added for transportation, insurance, and other charges, including duty on importation into this country. Surely no sane mind can impute this great and sudden rise in price of iron in England to the operation of the American tariff.

For the purposes of our comparison, we have in each instance taken the highest price of wool and of iron during the month quoted, as being the price most difficult to maintain against contact with depressing influences, and thus manifesting the largest purchasing power consistent with surrounding circumstances.

Months.	1853.			1869.			1854.			1870.		
	Highest prices of wool..........	Highest prices of common English bar iron..........	Pounds of wool for a ton of iron.	Highest prices of wool..........	Highest prices of common English bar iron..........	Pounds of wool for a ton of iron.....	Highest prices of wool..........	Highest prices of common English bar iron..........	Pounds of wool for a ton of iron.....	Highest prices of wool..........	Highest prices of common English bar iron..........	Pounds of wool for a ton of iron.....
January	$0.40	$70.00	175	$0.65	$90.00	138	$0.40	$70.00	175	$0.55	$80.00	145
February......	.44	73.00	165	.65	90.00	138	.38	70.00	184	.60	80.00	133
March.........	.44	75.00	170	.60	90.00	150	.38	72.50	190	.60	80.00	133
April.	.44	67.50	153	.6c	95.00	158	.38	75.00	197	.60	72.50	120
May............	.44	70.00	159	.68	9c.00	132	.38	77.50	203	.60	72.50	120
June	.44	63.00	143	.65	90.00	138	.35	72.00	206	.60	72.50	120
July............	.44	57.50	130	.60	90.00	150	.33	73.50	222	.56	72.50	129
August........	44	60.00	136	.60	90.00	150	.30	73.50	245	.56	80.00	142
September...	.44	67.50	153	.60	90.00	150	.30	73.50	245	.56	80.00	142
October	.44	67 50	153	.60	90.00	150	.28	73.50	262	.56	75.00	133
November...	.40	67.50	168	.60	90.00	150	.28	73.50	262	.56	80.00	142
December ...	.40	65.00	162	.55	90.00	163	.28	65.00	232	.56	80.00	142

Months.	1855.			1871.			1856.			1872.		
January......	$0.27	$60.00	222	$0.56	$72.50	129	$0.34	$61.00	179	$0.70	$110.00	157
February....	.27	60.00	222	.56	70.00	125	.34	61.00	179	.70	110.00	157
March.........	.27	60.00	222	56	70.00	125	.34	62.50	183	.70	115.00	164
April...........	.27	57.50	212	.56	70.00	125	.38	65.00	171	.90	102.50	113
May.......... .	.34	60.00	176	.56	70.00	125	.38	65.00	171	.85	100.00	117
June	.34	55.00	161	.60	70.00	116	.38	62.50	164	.85	112.50	131
July	.34	57.50	169	.70	72.50	103	.36	62.50	173	.80	110.00	137
August	.34	60.00	176	.70	72.50	103	.36	60.00	166	.80	....	.....
September...	.34	62.50	183	.70	72.50	103	.34	57.00	167	.65	105.00	161
October	.34	65.00	191	.70	72.50	103	.34	57.00	167	.60	105.00	175
November...	.34	65.00	191	.70	72.50	103	.37	60.00	162	.60	100.00	166
December ...	.34	57.50	169	.70	72.50	103	.37	65.00	175	.75	92.00	122

Months.	1857.			1873.			1858.			1874.		
January	$0.37	$57.50	155	$0.75	$90.00	120	$0.32	$55.00	171	$0.55	$80.00	145
February....	.37	58.00	156	.75	90.00	120	.32	55.00	171	.60	78.00	130
March.........	.44	62.00	140	.73	90.00	123	.32	55.00	171	.60	80.00	133
April..........	.44	62.50	142	.73	90.00	123	.32	5c.00	156	.60	80.00	133
May	.42	62.50	148	.55	90.00	163	.32	47.50	148	.60	78.00	130
June	.42	56.00	133	.55	100.00	181	.32	49.50	154	.65	78.00	120
July............	.38	52.00	136	.53	100.00	188	.32	46.50	145	.60	78.00	130
August........	.38	55.00	144	.53	95.00	179	.32	46.50	145	.60	78.00	130
September...	.38	54.00	142	53	85.00	160	.32	46.00	143	.62	75.00	120
October	.35	53.50	152	.55	85.00	154	.32	45.00	140	.63	72.00	114
November...	.35	53.50	152	52	82.50	158	.32	46 00	143	.65	75.00	115
December....	.35	53.50	152	.53	80.00	150	.32	47.00	146	.65	78.00	120

If any reader will take the trouble, as we have done, to check off, without regard to dates, the highest number of pounds of wool for a ton of iron under Protection, against the highest number under partial Free Trade, *he will find in every case that it required fewer pounds to make the purchase under Protection.* Recollect, this is no restricted contrast between scattered months or years, selected merely to make out a case, but a comparison between two periods, each of six consecutive years divided into months, under different systems of tariff. We ask, in all candor, appealing to common sense, whether the result of our investigation supports the Protectionists or the Free Traders? Can farmers believe they are duped, victimized, robbed by tariffs under which the purchasing power of their wool for bar iron has been so largely increased?

When we confine the comparisons to sets of two years, there is a similar showing. In the whole range of the exhibit for 1853 there are only two months in which a ton of iron could have been bought with fewer pounds of wool than in 1869. *In* 1870 *the highest number of pounds of wool required to make the exchange was less—considerably less—than the lowest number needful in* 1854. *The same fact is true in comparing* 1871 *with* 1855. In only three months of 1856 would nearly as few pounds of wool as in 1872 exchange for a ton of iron. The only showing in the whole statement in favor of partial Free Trade is in 1857—the panic year—when, for eight months out of the twelve, the purchasing power of wool was greater than in 1873, also a year of financial revulsion. Finally, comparing highest with highest purchasing power, there were only two months in 1858 when wool would pay for more iron than in 1874. These conclusive statistics demonstrate that the Free Traders are wrong, and leave no peg to hang a doubt upon.

It will be noticed that the prices of wool have been much *higher* under Protection than they were under partial Free Trade. This is due to our policy of Protective tariffs, which has created an active home demand for the domestic product, that not yet amounting to an adequate supply. During the four fiscal years ending June 30, 1861, under the low tariff of 1857, we exported 4,494,572 pounds of wool, valued at $1,194,782; during the four fiscal years ending June 30, 1875, we exported only 713,278 pounds, valued at $188,981 We have now a demand at home for the wool crop, and our farmers do not need to look abroad for purchasers. To show the sudden and vast growth of the wool industry and the woolen manufacture in the United States under Protection, we have compiled from the census reports the following tabular statement for three several periods.

Particulars.	1870.	1860.	1850.
No. of woolen mills	2,891	1,260	1,559
Sets of cards	8,366	3,209	....
Hands employed	80,053	42,728	27,682
Capital invested	$98,824,531	$30,862,654	$28,118,650
Wages paid	$26,877,575	$9,610,254	....
Wool consumed, lbs	172,078,919	83,608,468	70,862,829
Value of products	$155,405,358	$61,894,986	$43,207,545
No. of sheep	28,477,951	22,471,275	....
Domestic wool crop, lbs	100,102,387	60,264,913	....
Population	38,558,371	31,443,321	23,191,876

These significant figures tell the story of benefits conferred upon farmers by the Protective policy. What else but the demand created by it for domestic wool could have induced such a heavy increase in the number of sheep? What else could have kept the price of wool up to such high figures in the face of such a large addition to the supply? See how slowly the woolen manufacture crept forward between 1850 and 1860, the number of establishments absolutely diminishing—that is, the big fishes eating up the little ones—under the Free Trade tariffs of 1846 and 1857. Then observe the advance by mighty strides between 1860 and 1870, under tariff Protection, the progress having been very much faster than that of population. Of the wool consumed by the mills in 1870, as much as 154,767,095 pounds had been derived from various seasons of the domestic crop.

Farmers should study these statistics, which bristle all over with facts closely allied with their interests. To cap all, staple woolen goods are cheaper than they were before the war, as the prices current will prove. Yet Free Traders, with monstrous absurdity and dogmatic assurance, insist that Western farmers are plundered by our system of tariff. *They never were more prosperous, and they know it.* They, of all classes, were least affected by the panic of 1873, *because of Protection to home industry.*

CHAPTER VIII.

THE FARMERS' SALT QUESTION.

THE iniquity of the tariff on salt—the way the duty on salt robs all classes, particularly the farmer—the bounty which the duty extorts from every consumer of salt, merely in order to enrich that bloated monopolist, the salt manufacturer—these have been an unfailing resource for Free Trade orators and writers in appealing to the people against the Protective policy. So, in his recent contribution to the September number of the *Atlantic Monthly*, David A. Wells could not abstain from repeating the stale and spurious argument. We quote from it as follows :

> For a number of years subsequent to 1860, Congress, with a view of protecting the American producer, imposed such a duty on foreign salt as to restrict the import and at least double the price of this commodity, whether of foreign or domestic production, to the American consumer. The result was, taking the average price of No. 1 spring wheat for the same period in Chicago, that a farmer of the West, desirous of buying salt in that market, would have been obliged to give two bushels of wheat for a barrel of salt, which, without the tariff, he would have readily obtained for one bushel.

These averments relate to matters of fact, and the facts contradict Mr. Wells point blank. It is manifest that his statements are made at random. If the price of salt has been doubled under Protection, and a barrel of salt requires two bushels of wheat to pay for it, whereas one bushel sufficed under partial Free Trade, as he asserts, then the price of wheat must have remained stationary; for, if the price of wheat has advanced, less than two bushels of wheat would purchase the barrel of salt. Wheat having really increased in price, the position taken by Mr. Wells *must of necessity be false*. He also insists that the tariff has restricted importations. Here he is again refuted by the facts. For the four years ending June 30, 1861, with 15 per cent. duty on salt, we imported 2,632,-

551,880 pounds of salt, at an invoice value of $5,006,197; yet, for the four years ending June 30, 1874, we imported 2,963,204,738 pounds of salt, at an invoice value of $6,591,243, with Protective duties on salt. If this is restriction, what does restriction mean? According to the dictionary, that word signifies "confinement within bounds;" but, in this case we have instanced, the bounds have been broken through. Again, Mr. Wells speaks of an "*average* price of No. 1 spring wheat." Now, that statement involves an impossibility. To obtain an *average* price, the rule of arithmetic requires that the total value shall be divided by the aggregate quantity. As no such record of values and quantities has been preserved, it is beyond the power of man to arrive at the *average* price. When, therefore, Mr. Wells rests his argument upon an *average* price, he rests it upon an impracticable assumption—upon something that has no ascertainable foundation. Such are the theoretic follies of the Free Trade system.

Nevertheless, we shall pile evidence upon evidence against Mr. Wells. To begin with, we quote from an able article in the *Evening Journal*, of Chicago, July 27:

Let us now compare the actual facts with what Mr. Wells asserts are the facts. We have not the Chicago prices at hand, but we have a table of comparative prices, compiled by the Milwaukee *Evening Wisconsin* (a Free Trade paper), and as the prices of salt and wheat in Chicago and Milwaukee never vary to any great extent, these figures may be accepted as, to all intents and purposes, reliable:

	1860.	1872.	1873.	1875.
Gold	$1.00	$1.12	$1.10⅜	$1.11
No. 1 wheat	.80	1.29¼	1.20½	1.25
Salt, per barrel	1.90	2.40	1.90	1.40

In 1860, while the Free Trade tariff of 1857 was still in force, it required two and a half bushels of No. 1 wheat to purchase a barrel of salt, instead of one bushel, as Mr. Wells would have his readers believe. This is a very remarkable error of fact for a man who claims to be an original investigator, and whose opinions are quoted as law and gospel by the Free Trade fanatics; yet his language can bear no other construction than that one bushel of grain should have theoretically been equal to one barrel of salt, whereas, as a matter of fact, it actually required two and a half bushels of wheat to buy a barrel of salt. On the other hand, in 1873, under this tariff, which Mr. Wells condemns so freely, our Western farmers were enabled to purchase a barrel of salt for one and a half bushels of No. 1 wheat, so that there was an actual saving of just one bushel under the tariff, instead of a loss of one bushel, as Mr. Wells states.

But this is not all. The commercial columns of the same issue of the Chicago *Tribune* which contained Mr. Wells's great effort show that the people of the

Northwest can buy a barrel of salt for 1.12 bushels of wheat at the present time, in the Chicago market, under this awful tariff that Mr. Wells and his friends, the disinterested English manufacturers, find so much fault with. As compared with the Free Trade prices of 1860, our farmers are able to save 1.38 bushels of wheat on the purchase of every barrel of salt used upon the farm. This saving, according even to Free Trade logic, must be credited to Protection.

The chief priest of the American Free Trade synagogue has not only misrepresented the facts, but has had the audacity completely to reverse the facts; or else he is perfectly ignorant of the facts.

Conclusive as this answer is, we shall place it upon still broader and more impregnable ground. Accordingly, we compare the purchasing power of wheat for salt in the years 1872, 1873, and 1874, under Protection, with such purchasing power in the years 1856, 1857, and 1858, under partial Free Trade. The contrast between these two periods is peculiarly apt and cogent. Each embraces a year before a panic, a year with a panic, and a year after the panic. We take the prices of wheat and salt for the earlier period from the annual reviews of the trade and commerce of Chicago, for the several years, as published at the time in the *Daily Press and Tribune*, of this city, and the prices for the latter period from the official reports of the Chicago Board of Trade. In all cases the highest price in each month has been used, because the top of the market is always most sensitive to a downward tendency, and thus represents the greatest purchasing power of the commodity that can be maintained amid the surrounding circumstances. All the prices were the ruling ones for *cash*.

	1856.			1872.		
Months.	Salt, fine, per barrel............	Spring wheat per bushel...........	Bushels of wheat for a barrel of salt...............	Salt, fine, per barrel	No. 2 spring wheat per bushel..........	Bushels of wheat for a barrel of salt..............
January.........................	$2.37½	$1.35	1.7593	$2.40	$1.25	1.9200
February........................	2.51½	1.30	1.9346	2.20	1.26¼	1.7426
March............................	2.50	1.05	2.38c9	2.20	1.25¾	1.7495
April..............................	2.50	1.20	2.0833	2.15	1.35¾	1.5836
May.	2.12½	1.14	1.8640	2.10	1.60½	1.3084
June, ...	2.06	1.06	1.9434	2.10	1.55	1.3548
July	2.00	1.00	2.[illegible]00	1.85	1.32	1.4015
August	2.06	1.10	1.8727	2.00	1.61	1.2422
September	2.00	.98	2.0408	2.10	1.28¾	1.6311
October	2.06	1 08	1.9074	2.20	1.21	1.8182
November..........	2.06	.78	2.6410	2.25	1.11¼	2.0225
December.....................	1.95	.78	[illegible].5[illegible]00	2.40	1.21	1.9835

Months.	1857. Salt, fine, per barrel....	Spring wheat per bushel..........	Bushels of wheat for a barrel of salt............	1873. Salt, fine, per barrel..........	No. 2 spring wheat per bushel.................	Bushels of wheat for a barrel of salt..........
January........................	$1.95	$0.87	2.2414	$2.40	$1.26	1.9048
February........................	2.00	.90	2.2222	2.40	1.26½	1.8972
March........................	2.06	.91	2.2637	2.30	1.22¾	1.8737
April........................	2.06	.88	2.3409	2.25	1.25	1.8000
May........................	1.90	1.10	1.7273	2.25	1.34	1.6791
June........................	2.00	1.25	1.6000	2.15	1.28½	1.6732
July........................	......	1.27	....	2.00	1.23½	1.6194
August........................	1.75	1.14	1.5351	2.00	1.46	1.3697
September........................	1.85	.96	1.9271	2.00	1.20½	1.6598
October........................	1.85	.77	2.4026	2.00	1.09½	1.8265
November........................	1.90	.69	2.7536	1.90	1.09½	1.7352
December........................	2.00	.54	3.7037	1.90	1.16¾	1.6274

NOTE.—The report from which we quote does not give any quotation for fine salt during the month of July, 1857.

	1858.			1874.		
January........................	$1.90	$0.57	3.3333	$1.90	$1.26⅝	1.5005
February........................	1.90	.57	3.3333	1.90	1.24	1.5323
March........................	1.90	.61	3.1148	1.90	1.23½	1.5385
April........................	1.90	.62	3.0645	1.90	1.28	1.4844
May........................	1.70	.65	2.6154	1.80	1.27½	1.4118
June........................	1.60	.65	2.4614	1.70	1.23	1.3821
July........................	1.55	.68	2.2794	1.70	1.17½	1.4468
August........................	1.47	.72	2.0417	1.70	1.10	1.5455
September........................	1.48	.88	1.6818	1.60	.99¾	1.6040
October........................	1.55	.88	1.7614	1.60	.99½	1.6080
November........................	1.50	.75	2.0000	1.60	.92⅝	1.7274
December........................	1.60	.70	2.2857	1.65	.92⅞	1.7766

These authentic figures should put Mr. Wells to the blush; for they strip off the pompous disguise which has long hidden a statistical blunderer from the public gaze. He asserts that salt under Protection has doubled in price. Where is the evidence of that in the above quotations? He necessarily implies that the price of wheat has remained stationary. What support does this find in our tables? He insists that the barrel of salt which would have cost the farmer only one bushel of wheat under partial Free Trade now costs him two bushels. This is contradicted point-blank by the facts. In twenty-three out of thirty-six months, under Mr. Wells's pet tariff system, it required two or more bushels, and in five of those twenty-three above three bushels of wheat to purchase one barrel of salt; yet, in all the thirty-six months under Protection, there was only one in which so many as two bushels of wheat were required to pay for a barrel of salt. Moreover, in the Free Trade period there were only four months in which the stipulated buying could have been accomplished with less than 1¾ bushels of wheat; but, in the Protective period, it could have been done in twenty-six months, or in all except ten. These are facts, and they are facts which demolish the argument of Mr. Wells, and place him in a very awkward attitude before the people—that of attempting to obtain their convictions on false pretenses. Our comparative tables fully and triumphantly vindicate our present tariff system from the aspersions of its enemies, by conclusively showing *that the farmer's wheat will purchase more salt under Protection than it would under partial Free Trade.*

In its issue of August 7th, 1875, the Chicago *Tribune* quoted Saginaw, Onondaga, and Canada salt, fine, at $1.40 per barrel, with the lowest cash price of No. 1 spring wheat at $1.34 per bushel. On the basis of these quotations, 1.0448 bushels, or scarcely more than one bushel of wheat, would have purchased a barrel of salt. The lowest cash price of No. 2 spring wheat, same day, is given at $1.28½ per bushel, at which rate a barrel of salt could have been paid for with 1.0895 bushels of this quality. No such transaction could have taken place in the Free Trade years 1856, 1857, and 1858. Besides, at that period the farmers did not, as now, generally use labor-saving agricultural implements, nor did they have so many railroads or so cheap transportation. Making due allowance for these advantages, Western wheat has fully doubled its purchasing power in relation to salt, instead of decreasing fifty per cent., as Mr. Wells asserts by completely reversing the truth. After this humiliating disclosure, the Free Traders should stop their senseless clamor about the tariff on salt, and particularly about its robbery of the farmers.

CHAPTER IX.

THE FARMERS' LUMBER QUESTION.

FREE Traders, in organizing their campaign against the tariff, do not hope, and can not expect, to accomplish their intended end by a single sweeping blow. This necessitates attack upon a few particular interests at a time, the choice naturally falling upon those which touch human needs and uses at most numerous points, such as salt, iron, cottons and woolens, lumber, and the like. It has been represented to the farmer, over and over again, that he is the especial victim of Protective duties, and that those on lumber fall upon him with the pressure of a heavy yet unnecessary burden. All these allegations relate to matters of experience; hence the issue which has been raised can be settled only by an appeal to the facts. The real question is the purchasing power of farm produce. We have shown that wheat buys about twice as much salt under Protection as it did under partial Free Trade. Let us now see how much corn—another great staple of the West—has been required, under these opposite systems of tariff, to pay for a thousand feet of lumber.

Here we can not properly compare the three years 1872, 1873, and 1874 with the three years 1856, 1857, and 1858, as we did in the case of salt, because the great Chicago fire, Oct. 9, 1871, created a phenomenal demand, sudden and long-continued in its needs, for every kind of saw-mill product, at once advancing prices to unprecedented figures, maintained many months, and amounting to an almost immediate rise in value of 15.79 to 25 per cent. To accept such exceptional prices as a standard for comparison with other years, not complicated with extraordinary circumstances, would be as fallacious and as absurd as to take the price of cotton in New York, in 1864, with a range of 69 to 180 cents per pound, as a sample of the common level of quotations. In like manner we are excluded from employing the year 1858 in the calculation; for then lumber had dropped to figures unprecedentedly low. The *Daily Press and Tribune*, of this city, in its annual review of the trade and commerce of Chicago for that twelvemonth, said:

> The lumber trade during the past year has been very much depressed. The heavy shipments of 1857, being followed by a general stagnation of business, owing to the monetary crisis, left us on the 1st of January last with an immense stock on hand, and very little demand either from the interior or by the city. In the month of February, however, dealers saw the necessity of reducing the prices of lumber, and from that time up to the close of the year common lumber sold freely at $6@$8 per thousand feet. The manufacturers, however, did not recover from the depression, and not more than one-third of the amount sawed in 1857 was turned out during 1858. Nor indeed could they have done so with any advantage or profit to themselves, even had they cut the logs; for, at the prices which ruled here, unless the mills were economically run, manufacturers could scarcely clear expenses. As will be seen from the tables which follow, the receipts during 1858 were 186,618,692 feet less than they were in 1857.

Considering these peculiar surroundings of the case, we shall compare the years 1868, 1869 and 1870 with the years 1855, 1856 and 1857, in order, as far as practicable, to avoid exceptional elements of the problem to be considered. In some respects, these two periods offer an apt contrast, since each of them is removed to an almost equal distance from the beginning of a Protective era, on one hand, and of a Free Trade era, on the other; with this advantage against Protection, that prices of lumber in 1857—the panic year—were very much less than they had been during the preceding years.

The Annual Review of the Trade and Commerce of Chicago for 1858, published by the *Daily Press and Tribune*, gives the range of cargo prices for lumber, and the highest and lowest prices for corn on the first day of the month for five years. From that source we have derived our figures for the period of partial Free Trade. The quotations for the Protective period have been taken from the official reports of the Chicago Board of Trade. In each instance we have adopted the highest price as the one most sensitive to the depressing touch of surrounding circumstances, and as manifesting the greatest purchasing power that could be maintained during the month. It should be added that prices for 1855,

1856, and 1857 represent the best and costliest corn in the market, while the prices for 1868, 1869, and 1870 stand for No. 2 corn, with the exception of 1868, for which, on the first day of each month, the report gives quotations for No. 1 only. With these explanations, we direct attention to the comparative tables which follow.

Months.	1855. Highest prices of corn first day of month........	1855. Cargo prices—Highest per M feet..............	1855. Bushels of corn for M feet of lumber..........	1868. Highest prices of No. 1 corn first day of month....	1868. Cargo prices—Highest per M feet..............	1868. Bushels of corn for M feet of lumber...... ...
April.........	$0.55	$.......	....	$0.84	$22.00	26.1905
May	.69	15.00	21.7391	.88¼	22.50	25.4958
June	.76	14.00	18.4211	.85	18.00	21.1765
July	.73	14.50	19.8356	.86	17.50	20.3488
August	.72	15.50	21 5278	.98	17.00	17.3469
September.........................	.69	17.00	24 6377	1.00	17.50	17.5000
October	.64	17.00	26.5625	1.12	17.50	15.6250
November..........................	.72	16.00	22.2222	.80	17.00	21.2500
Months.	**1856.**			**1869.**		
April	$0.41	$	....	$0.55	$17.00	30.9091
May	.37	16.00	43.2432	.55½	16.00	28.8288
June	.33	14.00	42.4242	.59	16.50	27.9492
July................................	.41	14.50	35.3659	.70	15.50	22.1429
August.........................	.45	14.50	32.2222	.89	15.50	17 4157
September...........................	.38	15.00	39.4737	.88¼	16.00	18.0179
October.............................	.39	15.00	38.4615	.70½	15.50	21.9858
November	.31	16.00	51.6129	.65½	15.50	23.6641
Months.	**1857.**			**1870.**		
April..............	$0.37	$.......	....	$0.80	$15.25	19.0625
May	.60	13.00	21.6667	.87	15.50	17.8166
June	.72½	14.00	19.3103	.82¾	16.50	19.9396
July..............................	.66	15.00	22.7273	.81¾	16.50	20.1866
August............................	.68	14.00	20.5882	.84	17.25	20.5357
September.........................	.69	12.00	17.3913	.64¼	17.00	26.4591
October	.50	11.00	22.0000	.65	17.00	26.1538
November.........	.46	....	....	.55	16.50	30.0000

NOTE.—There are no quotations in the reports from which we have quoted for April in 1855 and 1856, nor for April and November in 1857 for lumber. In all cases the prices are for run of mill lumber.

We confidently submit these statistics as complete disproof of the Free Trade allegation that Western farmers are oppressed, victimized, and robbed by the tariff on lumber. During the Free Trade period the duty was only 15 per cent. ad valorem ; during the Protective period, and now, $2 per thousand feet. Yet what do we see ?

For seven months of the season of 1868, comparing highest price with highest price, lumber could have been bought with fewer bushels of corn than during the season of 1855. *In every month of the season of* 1869 *such purchase required less corn than it did in* 1856. Only in 1857—a year of crisis, panic, revulsion, and collapse—is there any showing in favor of the Free Trade policy. While a heavy decline took place in the prices of lumber in that year, the prices of corn were maintained nearly as high as they had been in 1855. Had the prices of lumber in 1857 been just what they were in 1856, the purchasing power of corn for lumber would have been generally less than it was in 1870. However, taking the results just as they stand, the folly and delusion of the Free Trade assertions become manifest; the more so when we consider that, according to the official reports of collectors of customs on our northern frontier, the Canadians, who are the only exporters of common timber and lumber to the United States, pay the duties out of their own pockets for the privilege of our markets, the entry charges thus falling upon foreigners, not upon American consumers, as is so often alleged. If any credit is to be given to the positive knowledge of experts, then the testimony of the collectors, almost daily brought into contact with the facts of the case, should be conclusive against the mere suppositions and theories of the Free Traders.

To complete our comparison, we turn to the quotations of the day. The Chicago *Tribune* gives 62¾ cents per bushel as the highest price for No. 2 corn on Sept. 2, 1875, and $14 as the highest price by the cargo for boards and strips. At these rates 22.3108 bushels of corn would pay for M feet of lumber—figures much lower than the general run of prices in 1855 and 1856. Nor should it be forgotten that the farmer not only raises corn with less cost to himself per bushel than he did in those Free Trade years, on account of much larger possession of labor-saving implements for the work of agriculture; and also that he realizes for his individual benefit a greater percentage than then of the current prices of his corn in Chicago, because he has transportation both cheaper and more extensive. In 1875, under a Protective tariff, fewer days' labor, either by the farmer, the mechanic, or the manual day laborer, will purchase a thousand feet of lumber. Such being unquestionably the case, the charge that consumers are oppressed and robbed by the duties on lumber becomes the veriest trash of nonsense.

CHAPTER X.

TARIFF DUTIES AND CONSUMERS.

FEW newspapers in the United States have been able to propagate so much error about duties on imports as the New York *Evening Post.* This is due mainly to its mode of discussing the question. As a rule, its editorial articles on the tariff embody scanty citations of individual facts, pertinent to the issue, and illustrative of the argument. Usually, the *Post* delights in general propositions which, for the most part, contain only a small fraction of the total truth, yet which are presented and reasoned on, not as fragmentary and incomplete, but as aggregate and entire. All the co-operative factors are assumed to be present in the statement which is taken as a basis for deduction; consequently, every progressive step of the dialectic method is an additional movement into the realms of error. A specific example of our meaning will be found in the following extract from the *Post,* Dec. 26, 1874:

> Tariff taxes do fall, must fall, as a rule, upon the consumers of the taxed goods. It is true that dealers sometimes, in order to tempt a brisker market, are temporarily willing to pay a part of the tax out of their own profits; but this can never be the permanent state of things. Trade is always carried on for the sake of the profits of it; these profits tend to reach an average level; and profits consequently will never steadily pay steady taxes levied on the goods by the sale of which the profits are realized. Such taxes are always ultimately thrown upon the ultimate consumers of the taxed goods.

To persons who have not dug down to the bedrock of such propositions and conclusions, those above are likely to appear reasonable or conclusive. So soon, however, as we test these propositions by existing facts, or by a long term of experience, we immediately detect their fallacy. For example, some weeks ago we proved in our columns, by the concurrent statements of four collectors of customs at the leading offices where the revenues are now collected on the northern frontier of the United States, that "*the import duty is paid by the Canada producer or manufacturer, and not by the American consumer;*" and that "*the same can be said in relation to grain, and in fact of nearly all importations*" of Canadian products into this country. This has been a fact for such a series of years that it has assumed the characteristics of *permanency.* Here, then, we have practice at war with the *Post's* theory, which sinks to the mean level of an unsupported assertion.

Or, take the initial proposition: "Tariff taxes do fall, must fall, as a rule, upon the consumers of the taxed goods." Let us consider this in relation to imports from European countries, and produce one of the children of experience upon the witness-stand. "Immediately before the construction of the first steel rail manufactory in this country foreign makers charged $150 per ton (equal then to $225 currency) for steel rails. As American works were built, foreign skilled labor introduced, home labor instructed, and domestic irons, clays, ganister, and spiegel (after many expensive trials) found to produce excellent rails, the price of the foreign article was gradually lowered, until it now (1870) stands at less than $79 per ton in gold, or $96.38 currency." So said a memorial by railroad managers to Congress asking for an increase of duty, in order to protect American manufacturers of steel rails against the crushing-out process of their foreign rivals in ruinously reducing prices. Notwithstanding the panic and consequent depression in railroad circles, we imported in fiscal year 1874 to the amount of 292,821,945 pounds of steel rails, or only 27,261,155 pounds less than in fiscal year 1873, at an aggregate invoice value of $9,771,175; equal to $74.75, gold, per ton. It thus appears that the competition for the sale of steel rails in the American market, created by production on our own soil, and sustained by the influences of tariff legislation, has resulted in cheapening prices to consumers more than 50 per cent., measured by a gold standard —a cheapening which covers the entire duty more than twice told. Had it not been for the competition undertaken and continued by our home producers, under the fostering care of our tariff, foreigners would have maintained their monopoly of our market, and no abatement from the makers' price of $150, gold, per ton, might have taken place. On this supposition our railroad companies, within the four fiscal years ending June 30, 1874, saved an outlay of $34,009,435.78; for, during that period, we imported 401,386 tons of steel rails, the whole of which quantity was retained for home consumption, at a total invoice value of $26,199,155.42, equal to an average price of $65.27, gold, per ton, showing a saving in coin expenditure of $84.73 per ton, or, in the aggregate, of $7,810,280.36 in gold more than the entire foreign cost of all the steel rails imported for our own use during the four years specified. Even admitting, for the sake of argument, that the above estimate involves an error of fifty or even of seventy-five per cent. too much, which can not be the case, still the result establishes, to a remarkable degree, the beneficial influences exerted by tariff legislation, Protective in character, in reducing prices to American consumers. It is evident, therefore, that the proposition of the *Post* leaves entirely out of view some of the essential elements of the effect of tariff taxes.

We might adduce various other examples of similar force. Why, we ask, are the instances of the cheapening of prices under Protective duties so very numerous, if they are merely exceptions to a general rule? And why is it that instances of commodities made dearer by Protective duties—instances which should outnumber the others more than a hundred to one—are not forthcoming? If they constitute the body of the facts why are they not put upon the witness stand and made to testify? We will answer: because they exist nowhere except upon paper; because they are merely speculative assumptions, not living realities. Were they actual circumstances they would be offered in evidence fast enough.

When the *Post* says that "Trade is always carried on for the sake of the profits of it," and proceeds to reason upon the proposition as if it embodied all the factors of the problem, that paper misrepresents the facts and commits a grave error. Let us illustrate this point. When farmer and miller are within easy reach of each other, they divide between them, on some equitable plan, all the flour made; but when considerable distance is interposed between the two, a third person, the transporter—in other words, a middleman—must be employed, who takes a share of the grain, or the money price of that share, to compensate him for his services in conveying the grain to the miller; and, again, a share of the flour, or the equivalent of that share, to pay him for his time and trouble in carrying the flour to the farmer, leaving less to be divided between the man who grows the grain and the man who converts it into flour. Ultimately, however, the miller might grind the transporter's share of the grain, taking therefrom his customary toll, and thus might secure for himself the same proportion of the whole quantity as if the transporter had not intervened; but the farmer must, in any event, suffer a positive and permanent loss. It is true, the farmer makes a gain by obtaining the conversion of his grain into flour; but, between his gain and that of the miller and the transporter, theirs not being complicated with a sacrifice, there is a large inequality of profitable result. Let this inequality be extended to a great variety and number of exchanges, covering the most of his purchases, then his impoverishment would be merely a question of time, or else his power of accumulation would be so seriously crippled as to prevent any considerable or rapid improvement of his condition. It thus appears that the circumstances of trade may be such that the *copious* gain will fall always to the share of one of the parties to the transaction, and the *scanty* gain invariably to the share of the other, all the aggrandizing tendencies being with the former, and all the depreciating tendencies with the latter. Consequently, when the *Post* says that "trade is always carried on for the sake of the profit of it," that paper states only a small part of the truth, and leaves altogether out of view the very important fact that there is a manner of conducting trade which inevitably results in an unequal and oppressive distribution of profit among the parties to such trade.

It is this injurious kind of trade that is advocated by the *Post*—a trade that is circuitous, foreign, abounding with middlemen, and making necessary a large use of the machinery of transportation. *External* commerce, or exchanges between different nations, is far less important and valuable to a country than *internal* commerce, or exchanges between its own inhabitants But the *Post* considers the foreign market the great consideration constantly to be kept in view, as if the infrequency of exchanges were preferable to their frequency, it being unavoidable that exchanges between parties distant from each other must be fewer than between parties near together, and that labor must have less employment in the former case than in the latter. The rapid circulation of commodities constitutes the material prosperity of national life, leading directly to mental development and moral improvement, and, as a necessary consequence, to a higher civilization; moreover, by the tendency of example, promoting the forces of societary excellence everywhere. To carry into practice the blundering precepts of the *Post* would be to put the country on the road to industrial ruin.

CHAPTER XI.

PRODUCTION AND FREE TRADE.

SALEM, Neb., Dec. 29, 1874.

To the Editor of the *Inter-Ocean:*

Being a constant reader of your paper, and seeing you are in favor of a Protective tariff, I would like more light on this subject. I am a farmer. It seems to me to be a plain fact that no portion of our country could be benefited by the stopping of our factories, and also plain that they will not stop unless they are likely to invest their means where it will pay better. Now, if Free Trade should stop them, will it pay better under that *regime* to produce? It seems to me that the whole question turns upon this point: What is the relative cost of manufacturing, say, cotton and wool, so as to add one hundred or any other given per cent. to the cost of production? The reason of my selecting these two is because they are the most profitable productions of the soil. Hoping you will deem these two questions worthy of consideration, I remain yours, for the right,

GEORGE WATKINS.

IF we fully comprehend our correspondent, his first proposition involves the common Free Trade fallacy that any manufacturer who finds his business a losing one can change his investment to some other form of production. In fact, he can do so only in exceptional cases. Take a woolen mill, for example. A capitalist puts money into a building, into machinery, into raw material, and, employing operatives, makes cloth. He discovers that his factory will not pay. How is he to change his investment? Is he likely to find a purchaser for a losing business? How is he to turn, without great sacrifice, if at all, to some other kind of industry? Perhaps his building might be devoted to different purposes; but what is he to do with his machinery, adapted only to convert wool into cloth? It is manifest that the man can not extricate himself without more or less loss, leaving him weakened financially, possibly bankrupt. If his mill stops running, what is to become of his force of working people, thrown out of employment? What effect will the stoppage have upon farmers in his neighborhood, and particularly upon those from whom he has been accustomed to derive his supply of wool? Finally, will not the mill-owner, after such a disheartening experience, feel averse to risking what remains of his capital in some other branch of manufacture? While, therefore, it is plain that no portion of our country could be benefited by the stopping of our factories, it is equally plain that the factories might by losses be forced to stop, and that the means of the proprietors, having taken the form of buildings, machinery and other fixed capital, could not be withdrawn to be invested where it would pay better, but would remain, for the most part, idle and profitless. A complete illustration of these conditions is to be found to-day throughout the West in the numerous flax mills, which can not be operated without loss, and which can not be sold at any price. These unremunerative flax mills are monuments of the baleful influences of that provision of the act of June 6, 1872, which removed all duty from jute butts, and thus adopted the principle of Free Trade regarding that article. Does it pay our farmers better under that *regime* to produce flax? Has not almost every flax grower experienced the blighting force of that Free Trade legislation?

The same class of injuries was extensively inflicted upon our people through the tariff of 1846, which substituted the policy of partial Free Trade for the policy of Protection to home industry. That act, although passed in July, did not go into operation until December. About three and and a half years afterward, or on May 15, 1850, Samuel Calvin, a representative in Congress from Pennsylvania, made the following undisputed statement on the floor of the House:

> The coal mines of our State, in which millions of capital have been invested, have been rendered unproductive, unprofitable. Some have been sold by the sheriff, others abandoned to dilapidation and ruin. *I am informed that the sheriff is the only man now making money in the great coal fields of Schuylkill County; and that the population of that county has been reduced about 4,000 within the last twelve or fourteen months. A large portion of our numerous iron establishments throughout the State,* I would say the larger portion of them, *have been broken up, sold by the sheriff, or have suspended;* and the little remnant are now sending up their daily petitions to us to save them from the ruin that must speedily overwhelm them also.

On August 12, 1850, Joseph Casey, another Representative from

Pennsylvania, made a speech on the floor of the House. The facts stated in the extract below have never been contradicted:

In the year 1846 there were employed in the State three hundred furnaces, with a capital of $12,000,000, producing annually up to 1847, 389,850 tons of pig metal. This was about the time the tariff of 1846 was enacted and was about to go into operation. In the two years succeeding that period—1848 and 1849—the amount of iron produced had fallen from nearly 400,000 tons to about 250,000 tons; and at the close of the present year it will have fallen down below 200,000 tons. Take in connection with this an additional fact: The whole history of the manufacture of iron in Pennsylvania shows that in a period of seventy-five years there have been erected 500 furnaces; and out of them 177 failures, or where they have been closed and sold out by the sheriff. Out of this 177 failures, *one hundred and twenty-four of them have occurred since the passage of the tariff of* 1846. Again, out of the three hundred blast furnaces in full operation when the tariff of 1846 was enacted into a law, *one hundred and fifty, or fully one-half, had stopped several months ago, and fully fifty more of those remaining are preparing to go out of blast with the end of the present season.*

It will be remarked that all these iron works were in successful operation, and that a profitable market existed under the operation of the tariff of 1842, and that, so far from any of them going out of blast, new ones were constantly springing into existence. The business was gradually rising into importance, and the consumption rapidly increasing. It was affording constant and profitable employment to the industrious and toiling laborer. But the protecting and fostering hand of the Government is removed, and we find in this brief period the disastrous change that has occurred.

Here, under an elaborate system of non-Protective duties, such as Free Traders consider desirable, we find the stoppage of a multitude of industries, with absolute loss of power to transfer the investments to any other branch of manufacture. Not only had the capital embarked in these coal mines and these furnaces become unproductive—it was crippled, prostrate, perishing. Whom did it pay better to produce under that *regime?* Certainly not the farmers; for they lost a regular market for a considerable part of their annual surplus, when thousands upon thousands of laborers lost employment and wages. How could any class of producers be rendered more active or prosperous by a scheme of legislation which sounded a death-knell in the ear of industry and enterprise? What benefit could accrue to the people at large from an act of Congress which resulted in depriving vast numbers of work, and reduced the pay of nearly all the rest?

The real question involved in the Protective policy is the question of employment and wages; it does not turn upon the relative cost of manufacturing, as between cotton and wool, or between one and another set of articles, as supposed by our correspondent. The laboring classes are the nation. They are the producers, and they are, moreover, the greatest consumers. Their expenditure makes the great home market. When an industrious population are employed, they not only enrich the whole community to the extent to which they themselves are enriched, but by the market which their prosperity affords to other industries. When the laborers of Chicago are in full employment, what a market they must afford, not only for the diversified products of the farm which can not be raised in a city, but even for their own productions! If, however, the people content themselves with producing the raw materials of the farm, the forest, and the mine, for export, to be exchanged for manufactured articles, then we hire foreign laborers to do the work of manufacturing. Our own laborers, being thus deprived of that work, and the industry of the country consisting in the production of raw materials, the only employment open to them is in that kind of production. It is manifest, in such case, that there must be a large increase in the production of raw materials, with much greater competition for their sale in a common market, and that a foreign one. Few exchanges can take place among persons who produce the same things, each having enough and a surplus, consequently such persons must look abroad for purchasers. The larger the quantity of the surplus seeking export, the stronger will be the competition for its sale, and the greater will be the tendency to a reduction of prices, since prices must go down when the supply is increased without increasing the demand. Such would be the condition of affairs under Free Trade. How can it possibly pay better under that *regime* to produce?

Every farmer should be a Protectionist, because the effect of the Protective policy is to increase the price of everything the farmer has to sell, and to reduce the price of everything the farmer has to buy. This proposition is not only confirmed by experience, but agrees with the well-known laws of demand and supply. The effect of the Protective policy, it is admitted on all hands, is to build up and increase the number of manufacturing establishments, and thereby to increase the demand for the raw materials and breadstuffs produced by the farmer, and thereby increase (not diminish, as Free Traders say,) the price of everything the farmer has to sell; and, by increasing the number of manufacturing establishments, increase the quantity of manufactured goods, and thereby reduce (not increase, as Free Traders say,) the price of the goods which the farmer has to purchase. Hence, by increasing the demand, you increase the price of everything the farmer has to sell; and, by augmenting the quantity of everything the farmer has to purchase, you reduce the price. Such is the well-known operation of the great law of demand and supply, universal and invariable in its results. Besides, by increasing manufactures, you withdraw a portion of the labor employed in agriculture and employ it in manufactures, making customers and consumers of those who before were rivals in the production of agricultural supplies. This transfer of labor from the farm to the factory has the double effect of diminishing the aggregate amount of crop surplus, while augmenting the demand for it, and of increasing the tctal of manufactures, thus augmenting the supply and decreasing the price.

Let agricultural and manufacturing industry flourish side by side, and you have everywhere occupation fit for all. There is appropriate employment for stolid strength, for manual skill and dexterity, for inventive genius, for the active and the sedentary, for childhood as well as youth and mature age—nay, even for decrepitude. The framework of industry becomes compact, self-supporting, all-embracing, knit, morticed, and clamped together. Markets are at home rather than abroad. Cost of transportation ceases to be a grinding, impoverishing tax. Prosperity reigns.

That system of tariff is best which most thoroughly diversifies industry, and which most fully supplies all classes of the population with regular employment and good wages. Such a tariff is Protective, and the farmer's friend. Free Trade is his great enemy.

CHAPTER XII.

BENEFITS OF TARIFF PROTECTION.

INDEPENDENCE, Warren Co., Ind., July 8, 1875.

To the Editor of the *Inter-Ocean:*

I see, by an answer given in last week's issue, you favor a Protective tariff, and claim that the masses are benefited thereby.

1. If you please, show me and many others if the Western farmers, mechanics, or merchants are benefited by it.

2. Does it not give the Eastern capitalist and manufacturer great advantage in selling their goods at higher prices?

3. Do not we, as consumers, eventually pay the duty on our own as well as foreign goods?

4. Has not Protection by high tariffs been the hue and cry ever since the early days of Henry Clay, and that by the Eastern manufacturer; and we, as old Whigs, were dragged into the Protective belief, because the great orator advocated that doctrine? OLD-FASHIONED REPUBLICAN.

ALL classes have been benefited by our Protective system. Under it the laborer has had steadier employment and higher wages, conferring larger purchasing power. Owing to this the merchant has been able to sell more goods, and to realize a greater aggregate of profits. Such increase of prosperity among those not engaged in agriculture has enabled them to buy and consume a more copious quantity of farm products, thus reducing the surplus, which, to obtain sale, *must seek a foreign market*, and helping to carry up the price by reducing the supply. More specifically, we may answer in the language of an editorial article in the Chicago *Evening Journal*, July 12, as follows:

Before a single cotton-mill existed in the United States, imported cotton cloth, of an inferior quality, sold for 22 cents a yard. When a Protective duty of 8 cents a yard was imposed, and cotton-mills built, the competition between the English and American manufacturers soon reduced the price of cloth to 7 cents a yard. So, too, before delaine-mills were built, imported delaines sold at 50 cents a yard, and in 1856 the competition between foreign and home manufacturers had reduced the price to 25 cents a yard, and under the present Protective tariff this competition between rival interests has reduced the price of delaines to 15 cents a yard. In black alpacas the same facts are apparent. In 1857 these goods sold for from 75 cents to $1.25 per yard. At that time all the American manufacturers imitated foreign trade marks in order to sell their goods. The tariff of 1861 and succeeding years stimulated the manufacture of alpaca, and to-day it sells at from 25 to 45 cents a yard, the quality being fully equal to the high-priced goods of 1857. The prices of cotton goods, coarse woolen goods, boots and shoes, hats and caps, iron and steel rails, and even bar-iron and salt are less to-day, in currency, than they were in gold in 1857, and it is pretty generally known that in 1857 prices were exceptionally low for partial Free Trade eras.

It requires only that the farmer should consult his memory to know that he sells his produce at higher prices, and buys his supplies more cheaply now than he did previous to the war. His agricultural implements are not only less than then in money cost, but they are of a higher grade of usefulness, are more durable, and accomplish their work in a more satisfactory manner. We appeal to the farmer whether he does not now enjoy larger comforts, have greater conveniences, realize higher profits, and see, generally, an easier time than he did previous to 1861. If this be so, has not the farmer been signally benefited by our system of Protective tariffs? Had the policy of Protection to home industry been hostile to the prosperity of the agricultural classes, the result must have made itself felt in the every-day life of the farmer. When we learn that he was never so thrifty and comfortably situated as at the present moment, we are forced to conclude that the policy of Protection has been, not only harmless to his interests, but positively advantageous.

Why should Protective duties reinforce the power of manufacturers, East or West, to demand high prices for their products? The necessary tendency of a Protective tariff is to increase the number of persons engaged in a particular branch of reproduction. This certainly means a more powerful and energetic competition for the sale of the fabrics made. Now we ask, in all fairness, does an addition to the persons who are producing a certain product operate to enlarge, or to reduce, their ability to compel the payment of higher prices? When more people are seeking to sell a given article, is the natural tendency of this competition for sale to increase or to decrease values? If a manufacturer has a constant apprehension that in the race for custom he may be undersold by a rival, is he thereby encouraged to put his price up? We put these questions to the common sense of the reader. There seems to be no logical conclusion other than that a tariff policy which leads to a multiplication of establishments in the same line of business must inevitably tend to a more intense competition for the sale of the product, and, through that, to a growing reduction of prices until the minimum shall have been reached.

The real cause of the hue and cry raised over the "enormous" duties on steel lies in the fact that the English producers are now compelled, by the exigencies of American competition, to pay a considerable part, perhaps all, of the duties out of their own pockets for the privilege of our markets. This we proceed to demonstrate. In 1874 we imported for home consumption, 8,738,483 pounds of high grade steel, valued at $1,094,222.32, the whole quantity having been *consigned* to *agents* of the European houses, not *sold* to American *importers*. These figures show an average invoice value of 12.522 cents per pound. Duties to the amount of $373,742.35 were collected on that steel. This shows an average duty per pound of 4.277 cents. Adding together the average invoice value and the average duty, we obtain the sum of 16.799 cents, *gold*. Now, the Boston *Journal of Commerce*, Jan. 30, 1875, quotes English tool steel at 17½ cents, *gold*, per pound. Next, subtracting the 16.799 cents from that price, we have a remainder of .701 of a cent., or scarcely more than 7-10ths of a cent in coin. If we allow reasonable estimates for ocean transportation, drayage, insurance, storage in the bonded warehouses, office expenses, compensation to agents, losses by bad debts, and so on, the seven mills must not only be swallowed up, but several cents besides, leaving absolutely nothing for profits. How is this deficiency to be made up—for it must be derived from somewhere—unless it is taken out of the pockets of the Sheffield manufacturers? Under all the circumstances, it is not possible to sell English tool steel in the American market at 17½ cents, gold, per pound, except on the assumption that the foreign producer, not the American consumer, pays the duty in great part, if not entirely. That is where the shoe pinches; for American tool steel, quite as good as the English, perhaps better, can be bought side by side with the foreign article in Boston, at 15 cents, currency, per pound. No wonder the duty is a burden—*to the Sheffield manufacturer!*

No error is more common or more unfounded, in discussing the tariff question, than the assumption made by Free Traders that the duty is added to the price, not only to the imported article which is dutied, but also to the home-made manufacture of like kind. This assumption is contradicted by a multitude of facts. In the first place, the duties on very nearly all imports from Canada into the United States, as the experienced collectors of customs on our northern frontier have officially declared, are paid by the Canadians for the privilege of sale in our markets. This one fact disproves the position taken by the advocates of an unrestricted commerce. Moreover, we have demonstrated in these columns, beyond room for doubt, that English manufacturers of steel are compelled to pay a considerable part, sometimes all, of the duty on that article before they can get it into our markets. Under these circum stances, how is it possible for the duty to be added to the price The present duty on wheat is twenty cents per bushel. Does any farmer ever take that rate into consideration in fixing his selling prices? Nevertheless, the tariff protects him, and assures him a higher quotation than he could possibly have without the tariff. Once repeal that duty, then Canadian wheat will pour into New England and New York, and there supply an annual consumption of some 32,000,000 bushels, now almost altogether furnished from the West. If our farmers were forced every year to throw that additional quantity upon the foreign market, or forego sale, would price tend up or down? Every tiller of the soil who thinks that the policy of Protection is injurious to his interests, and repressive of his prosperity, needs only to try five years or less of partial Free Trade to produce a permanent change of conviction.

CHAPTER XIII.

EFFECTS OF PROTECTIVE DUTIES.

OUR attention has been called to the following paragraph in the Chicago *Times*, of Nov. 18:

The advocates of the Protection piracy are just now busily engaged in demonstrating that tariff duties do not affect prices. If Uncle Sam puts a duty on any article, the foreigner forthwith drops his price, and his product is offered in our market at the same price as before. If that is so, two things must necessarily follow: First. Protection can not protect, and all those worthies who spend their time and money in Washington every winter, trying to induce Congress to raise the duties on their products, are laboring under a great delusion and spending their substance in vain. Second. Foreigners must give us some of their products out and out. The duties on many articles are equal to or greater than the foreign prices, and if the foreigner reduces his prices to the amount of the duties he must needs reduce them to zero, or a ruinous quantity, and actually pay us something for taking his goods. Such are some of the logical deductions from Protectionist doctrines.

The *Times* seems to be incapable of stating fairly an opponent's position. Now, the best evidence men can furnish of confidence in the accuracy of their own belief consists in frankly and honestly presenting the arguments of their opponents. In misrepresenting our views, the *Times* makes a virtual acknowledgment of the weakness of its cause.

Our proposition is that the general effect of *Protective* duties is to cheapen prices to consumers, not that "*tariff* duties do not *affect* prices." Indeed, in the very next sentence, the *Times* admits, in its blundering way, that such is not our position, by saying that "if Uncle Sam puts a duty on any article, the foreigner forthwith drops his price, and his product is offered in our market at the same price as before." The *Times* insists that Protection can not protect, if that be the fact. Fully to illustrate this point, we adopt a statement made many years ago by the Hon. Charles Hudson, a Representative in Congress from Massachusetts, as follows:

An article now *free from duty* is selling in our market for $1.20. The elements which make up this price are these: Cost in foreign market, $1; cost of importation, 10 cents; importer's profits, 10 cents: making $1.20. At this price the article can be manufactured and sold in this country. Now, let one of our citizens go into the manufacture of this article, and what will be the result? Why, the foreign manufacturer, who has heretofore enjoyed the monopoly of our market, and who is enjoying large profits, will immediately put the article at 90 cents to the American importer; this being the cost of the article. He will willingly forego all profit for the time being, for the purpose of crushing the infant establishment in this country; and the importer will give up one half of his profits rather than lose this portion of his business. This will reduce the price of the article to $1.05. The American manufacturer finds the article in the market at this reduced price, which is, in fact, less than he can manufacture the article for. He must, therefore, abandon his business, give up his establishment at great sacrifice, and yield the market to the foreign manufacturer, who, finding his rival destroyed, will immediately demand the old price, $1; and the consumer in this country will be compelled to pay $1.20, or perhaps $1.25, to make up the loss which the importer and manufacturer sustained during the period of competition. *This is the result when the article is free of duty.*

Now we will take the same article at the same price, both in Europe and America, *with a Protective duty.* A duty of 15 cents is imposed upon the article to encourage domestic manufactures. This, added to the former price, $1.20, brings the article up to $1.35. The foreigner fears the loss of the American market, and to prevent a surplus in his own market, and create a surplus here, he will at once put his article at cost, 90 cents. The importer will forego half of his profits, and take off 5 cents, which will bring the article down to $1.20, the very price which it brought before the duty was imposed. In the meantime the American manufacturer produces the article, which he can sell for the same price. Here, then, the manufacturer is protected, and the consumer has no additional price to pay. The importation will not be materially checked; and this, with the domestic production, will create a surplus, which will tend to a reduction of the price. A sharp competition will ensue, and necessity, that mother of invention, will bring out improvement in machinery; so that the article can be procured at a cheap rate. The skill also, which is acquired, will enable the manufacturer to turn off the article at less expense, and so afford it to the consumer at a reduced price. *Thus will discriminating duties protect the manufacture and cheapen the article.*

Such is the process by which "the foreign producer forthwith drops his price, and his product is offered in our market at the same price as before;" and such is the process by which prices are ultimately reduced to American consumers, while Protection is secured to home industry. Under Protective duties we gradually assume control of our own markets, drive the foreigner out, and supply domestic consumers with better articles at lower prices than before.

Let us illustrate this fact by an appeal to experience. The manufacture of axes and other edge tools was commenced at Hartford, Conn., in 1826, by the brothers Collins, who were the first to supply the markets of this country with cast-steel axes, ready ground for use. By the tariff of 1828 a Protective duty of 35 per cent. was levied upon imported axes. Under this Protection the Collins Co. introduced labor saving machinery, much of which was invented, patented, and constructed by themselves. Ultimately their axes altogether superseded the foreign article, on account of superior quality and cheapness. In 1836, foreign and home-made axes were selling side by side, in the American market, at $15 to $16 per dozen, at which time foreign producers, finding they could make no money at those rates, and that our establishments could not be broken down, withdrew from the competition, abandoning the entire market to our own manufacturers. Then home rivalry and improved methods continued the decline of prices. Axes were selling, in 1838, at $13 to $15.25 per dozen; in 1839, at the same; in 1840, at $13 to $14; in 1841, at $12 to $14; in 1842, at $11 to $14; in 1843, at $11 to $12; in 1844, at $11 to $11.50; in 1845, at $10.50 to $11; in 1846, at $10 to $11; in 1847, at $9.50 to $10 50; in 1848, at $8 to $10; and in 1849, at $8 to $10. These quotations are copied from the Finance Report of the United States for 1849, and they show a constant decline of prices, even after the pressure of foreign competition had been withdrawn. Now we are exporters of axes, and are wresting from the English one market after another. Said the Sheffield *Telegraph*, only a few weeks ago: "The steel of an American axe is so superior to that of an imported axe that no pioneer who understands his business will ever carry any other with him into the wilds." Such are the effects of Protective duties.

It is because of these results, often repeated, that British capitalists have employed their utmost energies, seconded by the diplomacy of England, to break down and crush out our rising industries as so many impediments to a monopoly control of the American market. This spirit is even now being exhibited toward the crockery manufacture in New Jersey, where labor-saving machinery is beginning to overmaster the competition of cheap manual processes abroad. A foreign manufacturer, who has been largely engaged in supplying the American trade, is said to have expressed a determination, before he returned to England last summer, to "unroof the Trenton potteries and destroy the 'plant' of capital there in the potting business." A British iron master said to a friend of ours, who was on a business visit to England in 1873: "When we get our grip on the throat of labor again, we will knock the bottom out of the Bessemer works in the States." Protective duties stand between such menaces and their intended fulfillment.

We have never maintained, as the *Times* intimates, that foreign producers *always* pay the duties, either in whole or in part. Sometimes they pay every cent, as the Canadians now do in almost every instance; sometimes, and this is oftenest the case, they pay part; and sometimes the entire burden falls upon the American consumer. These representations run parallel with long experience. The operation of the principle involved is thus stated by John Holmes, of Maine, in a speech delivered by him in the United States Senate, in 1832:

> If any one rule more than another is to be relied on, it is this: that, as soon as Protection begins to operate, and in proportion to its operation, the tax is reflected back from the consumer to the producer. Take the case of bar iron in the years 1818, 1826 and 1830, when the tariffs of 1816, 1824 and 1828 were in full operation. I recur to the price current in Boston, and select for an example "Old Sable." In 1818 the duty was $9 per ton, and the price, including the duty, $104. In 1826, duty, $18; price, including duty, $100. In 1830, duty, $22.40; price, including duty, $96. Thus, while the duty has been constantly increasing, the price of the article taxed has been as constantly diminishing. The reason is as manifest as the fact is true—the domestic article has been increasing in quantity. Suppose the foreign manufacturer furnished three-fourths of your consumption, the greater quantity would command the price, and this tax would fall on the consumer. But let the domestic product increase to one-half, the competition between foreign and domestic producers will be more equalized, and the tax will be divided between the producer and the consumer. Let the domestic product be three-fourths, and your own producer governs the whole market, and the foreign producer bears the tax or nearly so.

Our Protective system is the only plan of taxation by which foreigners, who do not bear any part of the burden of supporting our free institutions, can be made to contribute to the revenues of our Government, as an offset to the privilege of sale in our markets.

CHAPTER XIV.

PROTECTIVE TARIFFS AND PRICES.

FOR years the Chicago *Tribune* has discussed the tariff question on the dogmatic plan—assertion without proof. Not long ago that paper contained an editorial article under the title of "The Export Problem," in which a parcel of absurd allegations were presented as so many established facts. Our cotemporary should have learned by this time that it is far easier to make than it is to prove an assertion. For example:

The most material bearing that the American tariff has upon the export of American breadstuffs is, that since 1861 the American producer has received in exchange for his exports from one quarter to one third less in quantity in other commodities, such as iron, and cotton, and woolen clothing, than he did between 1846 and 1861.

Here, as usual, we have assertion without proof. The facts contradict the *Tribune.* Little more than three months ago a woolen manufacturer in Indiana, whose business was started in 1854, made the following statement in our columns—a statement derived from his personal experience:

But to show you just how cheap you are buying woolen goods (cotton goods will make nearly the same showing), I will give a table of prices in 1860 and 1874, simply for a contrast:

Choice tub wool, well washed, sold in 1860 for	$ .25 per ℔.
Average highest wages paid for hands in 1860	1.50 per day.
Price for 9-oz. jeans, wholesale	.60 per yard.
Tub wool, poorly washed, 1874, sold for	.50 per ℔.
Average highest wages paid in 1874	3.00 per day.
Price of 9-oz. jeans, wholesale	.50 per yard.

This exhibit emphatically denies the assertion made so dogmatically by the *Tribune.* The wool-grower obtains double the price, and the mill operative double the wages he did in 1860, and the wholesale price of the same class and grade of goods has declined 16⅔ per cent. Is that an illustration of the way in which Western farmers are literally *fleeced* by our Protective system, and forced by it to accept one-third less in quantity of woolen cloth in exchange for their produce than under the policy of partial Free Trade?

Take another example. High-grade English steel is to-day selling in the Chicago market at 21½ cents per pound, while the American article of equal quality is selling at 15 cents, both currency, and American manufacturers are supplying from three-fourths to seven-eighths of the home demand with increasing patronage, and are rapidly crowding their foreign competitors out of this country. Is that a further illustration of the way in which Western farmers are plundered by our Protective system, and compelled by it to take one-third or one-quarter less in pounds of steel in return for their breadstuffs than was the case under the tariffs of 1846 and 1857?

Here is still another example. Since the manufacture of porcelain has been established in the United States, the price of porcelain door-knobs at the factory has fallen from $12 to $3 per thousand, and the American article is now crowding Great Britain's product out of her colonial markets, where she has had a monopoly of the supply. Is that another illustration of the way in which Western farmers are robbed by our Protective system, and driven to take one-third less of porcelain wares in exchange for their grain than would have been the case between 1846 and 1861?

Next, let us turn to the phase of the subject presented by the following extract from the Finance Report of the United States for 1855, page 15:

Let it be considered that we manufacture all our furniture, all our carriages wagons, steam-engines, machinery for our factories and machine shops, most of our leather and shoes, boots, hats, door butts, and bolts of all descriptions, bells balances, buckles, brads, wood-saws, horse-cards, castors, curtain-pins, curtain bands, metal cocks, jack-screws, curry-combs, coal-hods, candlesticks, gas-fittings and burners, coffee-mills, caldrons, heavy edge-tools, hay and manure-forks, gimlets, hat and coat wardrobe hooks, harrows of all kinds, hoes, hollow-ware, planes, plows, sad-irons, tailors' irons, door-knobs, furniture-knobs, brass kettles, locks of all kinds, iron latches, lines, lanterns, lamps, levels, lead, cut-nails, clout nails, pins, pumps, punches, pokers, sand-paper, rulers, iron and copper rivets, ropes, rakes, oil-stones, wrought iron spikes, door-springs, window-springs, steelyards, scales of all descriptions, steel and brass scales, trowels of all descriptions, spoons of all descriptions, thermometers, tacks, vises of all descriptions, axes, wrenches of all descriptions, iron, brass, and copper wire, with a long list of other articles, to the exclusion of the like articles from other countries.

Since that day a multitude of other products has been added to this enumeration, some of which form a necessary part of every farmer's ordinary supplies. Now, we ask, what determines in our markets the current prices of manufactures of which we produce all that we consume, none being imported? What, unless it be cost in production and competition for sale? The inevitable tendency of rivalry in business is to reduce profits to a minimum, and to keep them there. Even Free Trade writers admit that. Consequently, the ever-recurring, never-ending competition among our domestic manufacturers for the possession of the same markets constitutes a force constantly moving in the direction of greater cheapness; for the moment any one should arbitrarily attempt to raise his prices, his rivals would step in with their lower charges and take away from him his customers. In what way can this rule of effects be injurious to Western farmers?

If it be answered that our Protective system operates to increase cost in production, and through that to enhance prices, we ask how such position is to be reconciled with the cases we have specified—cheaper woolens, cheaper steel, and cheaper door-knobs? If so, why do we now annually export increasing quantities of many finished products which found no export demand at all under the policy of partial Free Trade? If so, why did the proportion of manufactures in our domestic exports, in 1860, amount to 13 per cent. under partial Free Trade; yet, in 1874, amount to 19 per cent. under Protection? If so, why did we export, during the four years, 1858–61, of iron and steel and their manufactures, to the value of only $21,861,230, while, during the four years, 1871–74, we exported, of the same classes of articles, to the far greater value of $52,325,398? If so, why did we export, in the four years ended June 30, 1874, locomotives to the number of 247, and to the value of $3,590,648, yet not any under the tariff of 1857? As all these exports were destined to markets in which they encountered and overmastered that foreign competition which, it is said, could and would undersell our manufactures in our home markets, and thus force down prices, were it not for the restrictions imposed by our Protective tariffs, we wish to know how our producers are able to vanquish that foreign rivalry in foreign countries, away from the shelter of those tariffs?

That increased cost in production is not something peculiar to our Protective system is proved by the fact that cost in production of iron has largely augmented, in recent years, in Free Trade England, where there is no duty whatever on the imports of iron. For example, we imported from England, Scotland, and Ireland, in 1860, 157,602,032 pounds of pig iron, invoiced at $989,279 in the ports of departure. Those figures give an average invoice price of $14.06 per ton. In 1874 we imported from the same places 164,355,980 pounds of pig iron, invoiced at $2,386,726, equal to an average of $32.53 per ton. Here we see that the current price in the British market has increased 131 1/3 per cent., and that, too, under a system of entire Free Trade in iron. According to the process of reasoning employed by the Chicago *Tribune*, the only logical inference is that such great increase in price is wholly due to England's Free Trade policy. The corollary, by this method of drawing conclusions, is that, as the opposite system must produce opposite effects, the policy of Protection to home industry must diminish the prices of manufactured articles, pig iron among the rest. Thus is the *Tribune* entangled in the web of its own sophistry.

We wish to notice one more phase of the subject. In the Paterson *Daily Press*, April 14, 1875, we find the interesting and instructive statement which follows:

The Phœnix Mill, in fact, is one of the most wonderful of the many wonderful developments that have attended the progress of the silk industry in America, which has been so marvelous that we suppose nine out of ten cultivated Americans even yet do not know, and can hardly be made to believe without the evidence of their own eyes, that silks as perfect in dye and texture as are made anywhere in the world are now produced extensively in Paterson. Not many persons out of the trade know, either, that in many branches the American silks have driven the foreign fabrics out of the market, and of the latter scarcely any are now imported. This is notably the case with ribbons, and silks for ladies' kerchiefs, ties, bows, scarfs, and trimmings. One of the reasons of this is that the American manufacturers have wisely aimed at independence in all things, and do not depend upon the old countries now even for their patterns. The time was when the American market only received the "strippings," so to speak, of the foreign market. When London and Paris and other European markets were supplied with a favorite style of goods, what was left was sent to America, and thus became a stale thing by the time it got here. Now most of our American mills—like the Messrs. Tilt—keep their own designers, cut their own cards for Jacquard patterns, and are thus able to meet the demands of the market promptly, and to keep step with fashion in all her capricious and rapid movements.

This signal escape from slavish and disgraceful dependence upon foreigners is an outgrowth of our Protective system, without whose fostering encouragement the manufacture of silks would have had no successful commencement in this country.

We might take up and expose, one after another, the many errors of statement and fallacies of reasoning in the *Tribune's* article, but that is not necessary. With the above showing in hand, how can its utterances on the tariff question command the assent of intelligent people, or how can anything it says on the subject be trusted? The public mind no longer endures with patience anything like dictatorial instruction from the press, in schoolmaster fashion, saying, as to a parcel of fledgeling pupils: "Do you see this effect? Well, yonder is the cause." Nowadays, readers are satisfied only with severely logical processes and abundance of facts, so that they may themselves trace, step by step, the writer's argument, and determine for themselves its force and accuracy.

CHAPTER XV.

PROTECTIVE DUTIES AND PRICES.

CHICAGO, Ill., Feb. 11, 1875.

To the Editor of the *Inter-Ocean:*

A few days since you published an able article on the duties on tea and coffee, in which you advocated a continuance of those articles on the free list, because that policy would tend to *cheapen* them. How does this tally with your arguments that *Protective* duties tend to cheapen prices to the consumer? Please tell us.

A BELIEVER IN THE INTER-OCEAN.

A DUTY levied upon an article not produced, and not likely to be, or which can not be, in this country, can not be Protective in effect, because there is no present nor prospective existence on our soil of the industry involved. Tea and coffee belong to that category; hence duties thereon are not in any sense Protective of domestic labor and capital. Such duties are added to prices and are paid by consumers. As a general proposition, with very few exceptions, when such duties are repealed the prices at once recede by just so much.

But the case is different where we produce articles like those imported. Various effects follow, according to the character of the surrounding circumstances. If the duty, when diminished, was not sufficiently Protective, the reduction is equivalent to a free gift of the amount to the foreign manufacturer. A notable example of this took place in 1870, when the duty on pig iron was reduced from $9 to $7 per ton. The measure was carried through Congress on the plea that consumers would get their iron cheaper. But while the proposition was pending, and as soon as it had been ascertained that the measure would certainly pass both houses, the British manufacturers held a meeting, at which they resolved to put up the price of pig iron $2 per ton, or exactly the sum proposed to be taken off the duty. We find conclusive evidences of this movement in the statistics of imports officially published by our Government These evidences we present in the shape of the average invoice prices per ton, by months calculated from the import entries as given in the serial reports of the Bureau of Statistics, the ton being the custom house one of 2,240 pounds, as follows:

Months.	1869.	1870.	Months.	1870.	1871.
July	$17.52.1	$18.05.4	January	$15.61.5	$16.91.3
August	15.95.9	17.39.6	February	15.32.3	15.73.0
September	14.94.5	18.[illegible]8.7	March	16.00.4	18.53.7
October	16.35.0	17.31.7	April	15.74.6	16.06.7
November	16.95.8	17.79.6	May	17.65.9	17.56.5
December	15.32.1	16.70.2	June	18.16.7	17.73.7

The tariff bill became a law July 14, 1870, and the resolution of the foreign producers was adopted and carried into effect during the discussion in Congress. Now, in April, 1870, we imported at an average of $15.74.6 per ton; next month the price suddenly advanced to $17.65.9—an increase of $1.91.3, succeeded in June by a further rise of 50.8 cents. The general advance of $2 per ton was maintained for seven successive months, when the strength of American competition induced a fluctuating concession. December, 1870, was the last month under the $9 duty, and the average foreign price for that month was $16.70.2. In January, 1871—the first month under the $7 duty—the average was $16.91.3, and in June following, $17.73.7. Here we see no sudden decline of price, conforming to the reduction of duty, such as we pointed out in the cases of tea and coffee.

By the act of June 6, 1872, taking effect Aug. 1, the duty on pig ron was further reduced to $6.30 per ton. This was very inopportune and exceedingly foolish; for the $9 duty had been only moderately Protective, and the additional lowering of the tariff was made in the face of rapidly advancing prices abroad. England, the chief source of our foreign supply, had begun to feel severely the pressure of the coal deficiency and of the labor strikes, forcing a steady increase of cost in production—a fact made manifest by reference to the invoice prices. January, 1872, the average was $18.77.4; February, $20.75.6; March, $22.33.5; April, $23.03.5; May, $24.73.9; June, $26.69.9; July, $26.77.4; August, $29.26.2; September, $31.71.9; October, $33.39; November, $33.44.3; December,

$35.90.6. Under such circumstances, the seventy cents taken off the duty was a pure contribution out of the Treasury for the benefit of the foreign manufacturers. The reduction had no manner of effect upon prices; but the example exhibits the blundering ignorance which Congress sometimes applies to tariff legislation. Had the duty on pig iron been fixed at $12 per ton in 1861, and maintained at those figures, with corresponding rates on the metal in its higher forms, our iron industries would not be in their present half collapsed state, and thousands upon thousands of laborers, who to-day are compulsorily idle, would not have been without employment.

The tariff on steel was far more protective than that on pig iron, and there we find the cheapening effects of Protective duties. In the Boston *Journal of Commerce*, which is the organ of the Sheffield interest, English tool steel was quoted, January 30, at 17½ cents, *gold*, while American tool steel was quoted at 15 cents, *currency*, per pound. On that day the lowest price of gold was 113. Reducing 15 cents in paper money to its equivalent in coin at that rate, we obtain 13.275 cents as the result. The average invoice price of the highest grades of steel imported into this country in the fiscal year 1874 was 12½ cents, gold. Here we see, reckoning in specie, that American tool steel undersells the English article by 4.225 cents per pound, and is placed on our markets at merely a fraction of a cent higher than the average value charged in England, by the Sheffield manufacturers, in the sworn invoices of imports into the United States.

Within a week past the Chicago *Tribune* has repeatedly asserted that the duty is not only added to the price of the imported article, but also to the price of all of the home-made. If that wild allegation is true, why do not the American manufacturers of steel demand 17½ cents, gold, or its equivalent in currency? Their tool steel is quite as good as the English, and, for several purposes, far superior. Throughout the oil regions of Pennsylvania not a pound of foreign steel is used in boring wells; and there the resulting cost of breaking a drill is so considerable, sometimes involving the abandonment of the well itself, that price is nothing as compared with quality; so that the universal use of American steel in that broad area attests its supreme excellence. Then why do not our producers charge as much per pound as their English rivals?

It is because of constant competition among themselves to hold and to gain customers. If one of our manufacturers should arbitrarily undertake to raise his price, the rest would step in and take away from him his markets.

Briefly stated, the general effect of Protective duties is to arouse the activities of production and to provide work and wages for labor. When we buy abroad an article which might have been manufactured at home, we take away from our own mechanics to bestow upon foreigners the employment and pay for services involved in the fabrication of that article. If this plan of purchase is carried on extensively, the result is that thousands of our own people are deprived of opportunities to earn a livelihood in the arts of reproduction. The circle of occupations being thus contracted, there ensues a more energetic competition within that narrowed area for the sale of services, with the necessary consequence of diminishing wages and the laborer's purchasing power. Now, it is the ability of the great masses of the people to buy that creates universal prosperity. It is the expenditure of their earnings that causes the rapid circulation of commodities—the thrift of manufacturing establishments, the enterprise of merchants, transportation by rail and water, the growth of cities, the rise in the value of real estate, and the whole series of movements involved in material advancement.

CHAPTER XVI.

WHO PAYS THE DUTY?—THE PROPOSED CANADIAN RECIPROCITY TREATY CONSIDERED.

THE Canadians, it is said, sell us little of their wheat, and buy largely of ours. Then what causes the great anxiety to have our tariff taken off the importation of wheat? Our wheat and flour are admitted free *into* Canada. The abolition of our duties on the imports *from* Canada would not interfere with Canadian importations of grain from the United States. We have no export duty on our cereals, as Canada has on certain kinds of lumber, to be repealed by a reciprocity treaty, thus cheapening the invoice price. Now, if Canadian wheat and flour find *small* sales here, while ours obtain *large* sales there, why should such an ardent desire be manifested to have our duties on those articles abolished? We will answer. It is because the Canadians pay the duties out of their own pockets for the privilege of competing with our Western farmers in their markets, and thus are forced unwillingly to contribute to the support of our Government. Nor are we dealing in mere assertions. On this point the Collector of Customs at Plattsburg wrote, under date of June 8, 1868:

I submit the following statement as an illustration applicable very generally to all importations made into this district. The past spring large quantities of potatoes have been imported into this district, and *the duty of 25 cents per bushel, gold, paid by the Canadian seller or exporter*, as the sale has generally been perfected on the United States side of the boundary line, duties paid. The American speculator buying at such prices as to successfully compete with sellers in the Boston market, does the consumer of the imported potatoes pay the duty to the United States when he purchases the potatoes at the same price that another pays for the American product? If the Canadian can not export his goods at a profit, or the speculator can not buy in the foreign market and pay the duty at a price that he can sell at in the American market at a profit, he does not purchase. *The same can be said in relation to grain, and in fact of nearly all importations into this district.*

The Collector at Cleveland, Ohio, wrote, under date of October 20, 1868, as follows:

The chief articles of importation at this port are lumber and barley. The lumber market here is entirely controlled by the Saginaw market, and Canadian markets do not in the least influence us. The Canada market, to a great extent, is controlled by American markets, and the result is that the Canadian producer has to conform his prices to our market figures here; *this virtually makes the Canadian pay the duties on foreign merchandise imported here,* as he is compelled to sell his goods so as to enable the importer to pay the duties, and still not overshoot the American market. As the demand in Canada is not equal to the production, the producer is compelled to look to a foreign market for sale of his merchandise, and for this reason he must necessarily regulate his prices by that market to sell. The purchaser in buying always makes allowance for the duties, and the Canadian in his sales deducts the amount, and *thus in reality pays duty himself.*

The Collector at Oswego, N. Y., has this to say, under date of July 23, 1868:

The effect of the abrogation of the reciprocity treaty, in my opinion, has been the addition of several millions of dollars to the United States revenue at the expense of our Canadian friends.

There never appeared to me to be any true reciprocity in it, but rather the payment of a very large sum to them for something that was of very little benefit to us. As it now is, *the import duty is paid by the Canada producer or manufacturer, and not by the American consumer.* Any reduction in the rate of duties on importations from Canada would benefit them just as much, and would not lower the market value here.

The Collector at Buffalo, N. Y., makes the following statement, under date of Dec. 18, 1868:

The termination of the treaty of reciprocity between the United States and the Canadian Provinces, and the subsequent imposition of duties under the tariff enactments on articles of importation, has been a source of large revenue to the United States Government, *the burden of which has been borne by the foreign producer or manufacturer;* and any abatement or reduction of duties would, of course, redound to the advantage of such producer or manufacturer, and would not tend to reduce the value of the articles imported into this market.

These are the concurrent opinions of four collectors at the leading offices where the revenues are now collected on the northern frontier. They are the statements of men who are brought into daily contact with the realities of the case. All these eye-witnesses to the facts coincide in their testimony, and can not be mistaken. If experience is any test of knowledge, the Canadians pay the duties on imports from Canada into the United States. Now, what is proposed by the so-called reciprocity treaty? Why, to abolish the duties, and thus to transfer their burdens from the shoulders of *Canadian* farmers to the shoulders of *American* farmers, the great majority of whom are *Western* farmers, by reinforcing and strengthening the former, to the extent of the tariff amounts they have been paying out of their own pockets in their competition with the latter for the sale of grain in our markets. If Canadian farmers can afford to pay a tax of 20 cents, gold, equal now to 22 cents currency, on every bushel of wheat they export to the United States, then, were the duty abolished, they could add the 22 cents to their prices and sell just as many bushels on our soil as they do at present. Consequently, the reciprocity treaty would add 22 cents per bushel to the prices they now receive, and give that average sum to them as a free gift at the expense of our national revenues, the deficit having to be made up by imposing additional taxes upon other objects of taxation. But if Canadian farmers should conclude to overwhelm their Western competitors, then they could employ their gain of 22 cents per bushel in cutting down and overmastering the prices of Western wheat. In every such contest they would have an advantage of 22 cents per bushel more than they have now. Even under the tariff as it is, they sent into the State of New York alone no less than 5,649,798 bushels in the five years ended June 30, 1874. The Chicago *Tribune* has the folly to declare that this would not be injurious to the West, but a positive benefit.

Not long ago the *Ontario Reformer*, a representative journal of the Dominion, in discussing the question, "Do consumers pay duty?" used the following very decisive language:

The crop of wheat in the United States is officially estimated at 240,000,000 of bushels. We, as a Dominion, imported more wheat and flour than we exported in 1872, as per our government official returns. It is, therefore, very evident that we could not influence in the least degree the market price of wheat in the United States, and that *if we send our wheat there we lose the duty*. The proportion of our *surplus* of horses, cattle, sheep, and wool to the amount they consume is so very small that it is equally plain that we can not influence the price in their market, and that *we lose the duty*. The Americans consumed last year nearly 40,000,000 bushels of barley, of which we gave about one-tenth. If one-tenth can control the market price, then we can dictate the price of barley in the United States, and compel the consumer to pay the duty. We think that *our farmers lose the duty on barley*, or at least the greater part of it. The American people north of the Ohio consume not less than 8,000,000,000 feet of pine lumber per annum, of which we gave them not to exceed 700,000,000 in any year, or about one-eleventh. The city of Chicago alone annually receives more lumber than we export to all countries. We supply a large proportion of the peas consumed in the United States, and we think that the consumer of them pays the duty, but *this is the only natural product, whether from the farm, forest, mine, or sea, which we export to the United States in such quantities as will enable us to compel the consumer to pay the duty.*

Here the Canadian newspaper fully coincides in its statements with the statements of the four collectors. That print would not have dared to express such views among a people aware of the true state of the facts had the views been false. Its position, however, is fortified by a broad general principle, to wit: "*The man who must go to market is always compelled to pay the cost of getting there*, let it take what form it may, whether of freight, insurance, or charges at the custom house."

We will add the testimony of W. Martin Jones, United States Consul at Clifton, who wrote to the Treasury Department at Washington, under the date of Dec. 28, 1866, as follows:

The amount of exports, with the exception of lumber, from the Provinces to the United States, can have little effect upon the markets in the latter country, and the result is that THE DUTY PAID ON SUCH EXPORTS IS BORNE WHOLLY BY THE PRODUCERS, who, in receiving the benefits of the markets of the country, *are thereby compelled to bear a portion of the burden contributing to the support of its institutions.*

In this way the people of the Dominion annually contribute out of their own pockets from eight to ten millions of dollars toward defraying our national expenses. The Chicago *Times* talks as if this payment of part of our taxes by the Canadians were a great outrage and oppression upon American citizens, who should be allowed themselves to shoulder the burden of taxation.

CHAPTER XVII.

OUR MANUFACTURES GOING ABROAD.

OUR Protective policy has been so long and so comprehensively maintained, that we are beginning to reap some signal advantages which are merely the harbingers of what is to come, unless Congress should be so foolish as to stop the effect by discontinuing the cause. Only a few weeks ago the very significant paragraph which follows appeared in a newspaper published in Sheffield, England :

> We have been favored by a Sheffield merchant with an inspection of a number of saws made by Henry Disston, which have been sent here as samples. From what we have seen, we think it very desirable that specimens of American workmanship of this high class should be placed in the Museum at Weston Park, in order that workmen in the saw and edge tool trades may see with their own eyes and handle with their own hands such very tangible and instructing facts in steel. We have written over and over again, in order to present as vividly as was within our power, THE DANGER THAT SHEFFIELD WORKMEN HAVE TO FACE FROM THE COMPETITION OF OUR NOVELTY-LOVING AND ACUTE COUSINS OF THE NEW WORLD; and a few cases of American manufacture, if even temporarily placed on view, would probably serve to satisfy even the workmen of the saw trade that the policy of dislike to innovation and obstruction to changes involving improvements in production has been a most serious mistake.

While these admissions are very strong, they omit the very important fact that those saws made by Henry Disston & Sons—part of the Pennsylvania manufacturers whom the *Times* reviles—are offered for sale in the Sheffield market, England, at 15 per cent. less than the prices current there for the same class of steel goods. This competition at their very doors has filled the Sheffield manufacturers with consternation. Even the announcement of the intended consignment created a great stir in manufacturing circles. The October (1874) number of *The British Trade Journal* thus heralded the coming event in its "Sheffield report":

It is certainly a fact, and by no means a cheering one, that the orders from America for Sheffield goods are much lighter this season than they have ever been. American makers of files, saws, and tools are beginning to make their competition keenly felt. "Harry Disston," of Philadelphia, is again to the fore. He is said to be bent on teaching Sheffield saw makers rather a rough lesson. Not content with taking a good deal of American trade, he is now carrying the fire into the enemy's camp. We are told that his brother-in-law is on his way to England with a magnificent set of saw and tool samples, "resolved to wrest the home trade" from our townsmen. "Harry" employs 1,000 men, all non-unionists, and uses most novel and excellent machinery, by which it is claimed he can make saws AS GOOD as Messrs. Spear & Jackson's, and VERY MUCH CHEAPER. Sheffield saw makers will wait to see these famous samples with some curiosity. There is no doubt, however, that saws have long ceased to be a specialty of the Sheffield trade which defies competition. The Sheffield manufacturer is being roughly jostled in nearly every market of the world. At the same time, the end of the Sheffield saw trade is not yet at hand, despite that terrible Harry Disston, of Philadelphia.

This bold invasion of the British home market by American competition is only the image of what has been going on, step by step, in other foreign countries for some years. In the early part of 1871 the London *Times* editorially expressed the following admissions, which are exceedingly candid, pointed, significant, and impressive:

At this moment Birmingham is losing its old markets. A few years ago it used to supply the United States largely with edge tools, farm implements, and various smaller wares. *It does so no longer*, nor is the cause to be sought *merely* in the American tariff. It is found that the manufacturers of America actually superseded us, not only in their own but in foreign markets and in our own colonies, and the Birmingham Chamber has the sagacity to discover, and the courage to declare, that *this is owing to the superiority of American goods.*

High as are the wages of an English artisan, those of an American artisan are higher still, and yet the manufacturers of the United States can import iron and steel from this country at a heavy duty, work up the metal by highly-paid labor, and beat us out of the market, after all, with the manufactured articles. How is that to be explained?

The Americans succeed in supplanting us by novelty of construction and excellency of make. *They do not attempt to undersell us in the mere matter of price.* Our goods may still be the cheapest, but they are no longer the best, and in the country where an axe, for instance, is an indispensable instrument, *the best article is the cheapest, whatever it may cost.* Settlers and emigrants soon find this out, and they *have* found it out to the prejudice of Birmingham trade.

It thus appears that the greater intelligence and skill of American workmen, coupled with the greater inventive genius of our people, have been stimulated into the utmost activity under the fostering influences of our Protective system, so that we are wresting away from our formidable rival, England, one market after another, and are even threatening her supremacy upon her own soil.

We are falsely told that "with a strictly revenue tariff" our manufacturers would be enabled "to obtain control of nearly every market in the Western Hemisphere." Such is not the voice of experience. Every attempt we have made to develop our industries under that kind of tariff has ended in an illumination of suffering, and forced a return to the Protective policy. Instead of acquiring more and new markets, we began rapidly to lose control of our own at home. In urging upon us a reciprocity treaty and a tariff for revenue only, Great Britain is doing her utmost to regain her lost ground in this country.

CHAPTER XVIII.

FARMERS TAXED TO DEATH.

DURING 1871 the New York Free Trade League distributed all through the West immense numbers of an illustrated sheet, entitled "The People's Pictorial Tax-payer," in which appeared a statement, profusely elucidated with wood-cuts, to the effect that "The farmer rises in the morning, puts on his flannel shirt taxed 65 per cent.; and his trousers, taxed 60 per cent.; his vest, taxed 60 per cent.; and his overcoat, taxed from 40 to 150 per cent.; draws on his boots, taxed 35 per cent.; puts some coal, taxed 60 per cent., in his stove, taxed 55 per cent.; sits down to his breakfast from a plate taxed 45 per cent.; seasons his food with salt taxed 108 per cent."—and so on, until the poor fellow, taxed to death, sinks to rest in a grave covered with a heavily taxed tombstone, and slumbers where tariffs molest and vex no more.

Notwithstanding that the only truth in the whole representation was that "the farmer rises in the morning," all the other statements being false, this caricature of the facts was spread broadcast among our agricultural classes, as incontrovertible evidence that they symbolized Issachar—"a strong ass, couching down between two burdens"—all on account of our iniquitous and oppressive tariff system. The Free Trade press took up the cry, foremost among which was the Chicago *Tribune*, and repeated it to our farmers, in a multitude of forms from week to week. Tillers of the soil were everywhere told, with due emphasis, in the words of the Hon. Horatio C. Burchard, that "a double burden falls upon those the value of whose products must be measured by the price in the foreign market;" that "the enhanced price occasioned by the duty bears directly upon them as consumers, and ultimately they pay a portion of the tax imposed upon articles consumed by non-protected trades whose services they require;" and that "the cotton planter and wheat and corn grower pay the duty not only on their own clothing, wares, and implements, but also the increased price upon the goods and tools used by the carpenter, blacksmith, and workmen they employ." These misrepresentations were made in the face of the fact that Canadians pay the duties, amounting to eight or ten million dollars annually, on nearly everything they export into the United States; and that other foreigners either share with the importer the payment of the duty, or sometimes pay it all, the cases not being numerous where the entire duty falls upon the American consumer. For all that, falsehood was heaped upon falsehood, and repetition hastened on with the work of reiteration, until Western farmers were aroused into solicitude, urged into scrutiny of their surroundings, and awakened to a sense of oppression. After due search, they discovered this oppression, not in our system of Protection to home industry, but where it really was—*in the grinding tax of transportation to distant markets, and in the cost of maintaining a vast number of supernumerary middlemen.* Our farmers have learned that he who *must* go to a foreign market for the sale of his surplus *must* himself pay the cost of getting there, be that cost what it may. The granger movement for a redress of grievances has been the consequence—an outcome neither intended nor expected by the Chicago *Tribune* and its Free Trade co-laborers as the result of their false teachings.

Since the day when the crusade against our tariff system was organized, and "The People's Pictorial Tax-payer" was so widely distributed, we have had a financial crisis, and a universal depression of industry. Fortunately, no part of the United States has suffered so little from the panic as the West, and no part of its population less than the agricultural classes. These facts have so deeply impressed the Chicago *Tribune* that it felt impelled, in its leading editorial article, Nov. 23, 1874, to say what follows:

In the general talk at the East of hard times and depression of business, accompanied by the closing of mills and the reduction of working time in others, it is a comfort to turn to the more cheering figures which indicate THE PROSPERITY OF THE WEST. Last winter, after the panic had stricken the general business of transportation, the rates on grain were so much lower that, notwithstanding the close of navigation, the movement of grain was so heavy during the whole winter that the surplus of the Northwest standing over in the spring was very much reduced. So great was the reduction that, at the opening of navigation in the spring, the ordinary high rates of freight on the lakes for moving the winter accumulation did not prevail. The Western producers, therefore, began the season of 1874 with more cash in hand, all received during the winter, than had ever been known in any previous year.

If there is any truth in what the Chicago *Tribune* has been preaching for years, how comes it that our farmers, double-taxed by the operations of an iniquitous tariff system, ground down under foot by the bounties they were compelled to pay to enrich manufacturing capitalists, and more oppressed by Protective duties than any other class in the country, stand forth conspicuously as the most prosperous part of the population of the most prosperous section, when the blighting effects of panic and crisis descend upon the whole land? Is not this a very strange and contradictory result of carrying the chief burdens of tariff taxation for more than thirteen years? The manufacturing capitalists of the East, whose profits, so the Chicago *Tribune* insists, have been for a long period made plethoric by constant and large exactions wrested from our agricultural classes through our custom-house laws, are the greatest sufferers, while their victims thrive, least of all classes feeling the depression of business. All this is the reverse of what should have taken place, considered from the *Tribune* point of view. According to deductive consequence, the manufacturing capitalists should have been at the top of the heap of ruins created by the panic, while their dupes and victims, the farmers, should have been at the bottom. There is no reasonable explanation of this contradiction between teaching and facts, except that the *Tribune* has been the champion of a false theory of tariff, and has substituted misrepresentations of the truth for the truth itself.

In reality, our farmers are now just beginning to reap some of the substantial benefits of a long-continued policy of Protection to home industry. "Of all the pursuits of man," Mr. Carey says, "the last developed is a scientific agriculture." Its development *must be preceded by extensively diversified industry.* If we wish to find the scientific farmer, we must look for him amid a teeming population, and surrounded by a multitudinous development, material, intellectual, moral, and æsthetic, such as the Protective policy confers; for there alone can he secure those accessories which enable him to repay, promptly and regularly, to the land the vegetative constituents abstracted by the processes of tillage—constituents lost permanently to the soil by the butchery of a rude and ignorant cultivation. A perfected agriculture is the tardy product of a long, laborious, and extensive course of observation, experiment, and experience, looking to the use of the forces of nature for practical ends. No other one pursuit calls to its aid such a diversity of knowledge. The whole circle of the sciences and the arts is made tributary to its successful prosecution; yet a country devoted to the production of provisions, breadstuffs and raw materials—all the surplus being for export to foreign countries—can not possess, in an advanced state, the sciences and auxiliary arts most essential to its own industry. Thus, chemistry is indispensable to a prosperous agriculture; but who would expect to find that science, in its highest cultivation, in a community merely of farmers and herdsmen? We can not have a few isolated, solitary arts in complete excellence. They are social and gregarious. Each, in order to its success, requires the near and ready assistance of a hundred others. Only a manufacturing people can develop and sustain that diversity of the arts and the sciences which culminates in and is inseparable from a scientific agriculture. To the same extent that diversification enters the domain of labor, we may expect to see improvement in the methods and results of tillage. Carey says, and it is true, that "of all people, the last emancipated are the laborers in the field." Hence *Protection to home industry is emphatically the farmer's question.* He is interested in it more than anybody else.

CHAPTER XIX.

RAILROAD IRON—THE TARIFF—TRANSPORTATION.

LENORA, Fillmore County, Minn., April 8, 1875.

To the Editor of the *Inter-Ocean:*

Is it a fact that "Cheap Transportation" by means of road-bed, rails, and rolling stock, swollen in cost by a high tariff, is an impossible thing? Could you publish tables under partial Free Trade and Protective tariff bearing on this point?

J. M. WHEAT.

THE main question put by our correspondent was very pointedly and emphatically answered by Governor Carpenter, of Iowa, in his inaugural address, delivered Jan. 27, 1874, as follows:

Nor is it the tariff that burdens the farmer. An ingenious writer has shown, by estimating with great care and by unmistakable mathematical value and exactness, that if you take the New York Central Railroad and assume that it extends from Chicago to New York, double-track the whole distance, laid with iron weighing sixty-five pounds to the yard, and then assume that this iron represents only half of the road's consumption of iron, and further assume that the original cost of all this iron was increased by the entire tariff which would have been collected on each ton had it been imported—when he has granted all this and assumed all this, he demonstrates by actual computation, taking the cost of transport of one thousand and twenty-one million tons of freight, the amount this road carried one mile last year, that the exact additional charge on a bushel of wheat from Chicago to New York would be one cent and one hundred and eighty-eight thousandths of a cent on account of the tariff. The tariff will never ruin the Western farmer.

With such a state of facts, it is plain that road-bed, rails and rolling-stock *can not* be so swollen in cost by the present tariff as to make cheap transportation an impossible thing.

An indirect answer to our correspondent's question, with an argumentative force much stronger, is found in a memorial to Congress some years ago, signed by more than ninety officers and managers of leading railroads in all parts of the country, from Boston to Charleston, Milwaukee and St. Louis. These roads were and are large consumers of steel rails, yet they asked a duty of *two cents per pound*, equal to $44.80 per ton, on the imported rail, and offered the following reasons in support of their appeal:

Immediately before the construction of the first steel-rail manufactory in this country, foreign makers charged $150 per ton (equal then to $225 currency) for steel rails. As American works were built, foreign skilled labor introduced, home labor instructed, and domestic irons, clays, ganister and spiegel (after many and expensive trials) found to produce excellent rails, the price of the foreign article was gradually lowered, until it now (1870) stands at less than $79 per ton in gold, or $96.38 currency. Now that several millions of dollars have been expended in machinery, furnaces, and experiments in perfecting the process of manufacture in this country, and numbers of our own citizens are dependent upon it for support, the business is threatened with annihilation by the pressure of English and Prussian makers. We, as users of steel rails, and transporters of the food and material for American manufacturers and their numerous employees and skilled laborers, do not desire to be dependent exclusively upon the foreign supply, and therefore join in asking that, instead of the present *ad valorem* duty, a specific duty of two cents per pound be placed upon this article, being the rate fixed by a bill which passed the Senate Jan. 31, 1867, and of a bill which was reported to the House by the Committee of Ways and Means during the same year.

The eminent railroad managers, signers of the above memorial, represented about half the total length of all the railroad tracks in the United States, or some 26,449 miles. Now, if the inevitable effect of a high tariff is to increase the price of the dutied article, whether it be imported or home-made, as Free Traders assert, is it supposable that these experts in railway management, who annually had to buy and pay for thousands upon thousands of tons of steel rails, were ignorant of the fact? How could they have avoided knowledge of a fact with which they were brought in constant contact? With such premises, we are driven to the preposterous conclusion that more than ninety of the most experienced, sagacious and influential railroad men in this country deliberately appealed to Congress to so legislate as needlessly to increase the pecuniary outlay of their several companies in extending or in relaying their lines of track. What possible inducement could these memorialists have had thus literally to throw away money? When have railroad corporations been solicitous to be taxed and to pay taxes? Who ever yet saw such a spectacle? The very idea is absurd. It is incredible, because it is false. The memorialists knew perfectly well what they were about. *They were after cheaper rails.* Moreover, they were aware that the way to that end was through an increase of the duty. Although Congress did not grant the full measure of relief asked, the tariff on steel rails was changed from an *ad valorem* of 45 per cent. to a specific of 1¼ cents per pound, amounting in practice to an addition of 4 per cent. to the former rate during the period from Jan. 1, 1871, to Aug. 1, 1872. Under the stimulus imparted by this legislative encouragement and protection, coupled with the great demand for rails created by the rapid extension of our railroad system, various new rail works were established in the United States, by which a strong competition was engendered among American producers, besides the rivalry with foreign makers, the inevitable tendency being to lower prices. In short, the wise forecast of the memorialists has been made manifest in the result; for American steel rails are now selling at $75 currency per ton,* below the price realized for iron rails two years ago, and that, too, in our markets, where English steel rails can not be afforded at less than $95 currency. As a consequence, railroads may now be constructed at a lower cost than before, which is proof positive that our tariff, so far from being an impediment, is really an energetic aid to cheap transportation. Here, if it be true, as Free Traders assert, that the duty is necessarily added to the price of the dutied article, whether imported or home-made, the pertinent inquiry arises, Why are American steel rails sold at $75, when English steel rails are held at $95? Why do not domestic producers compel consumers to pay either the whole or part of that additional $20 per ton, when, according to Free Trade theory, they have full power to make that exaction? A theory thus contradicted by every-day experience must be worthless for any practical purpose, and certainly should not be adopted as a trusty guide in the enactment of tariff laws. These two *facts* are of more value and importance than a dozen cart-loads of Free Trade *theories*—that the price of foreign steel rails has been reduced, within a few years, by the pressure of American competition, from $150, gold, per ton, to $95, currency, and that our domestic makers are underselling the latter price by $20. Now, when it is considered that such competition, so potential in bringing down prices, has been the offspring of our Protective tariff, it must be admitted that the policy of Protection is conducive to cheap transportation.

*Since this chapter was written the price has been still further reduced.

A very high Free Trade authority, no less than David A. Wells himself, is officially committed to the same general view we have taken of the cheapening tendency and effects of Protective duties. In his report for 1867, as Special Commissioner of the Revenue, he said:

On steel much higher rates of duty than those recommended upon iron are submitted. Although these rates seem much higher, and are protested against by not a few American consumers of steel, yet the evidence presented to the Commissioner tends to establish the fact that if any less are granted the development of a most important and desirable branch of domestic industry will, owing to the present currency derangement and the high price and scarcity of skilled labor, be arrested, if not entirely prostrated. This is claimed to be more especially true in regard to steel of the higher grades or qualities. It is also represented to the Commissioner that, since the introduction of the manufacture of these grades of steel in the United States, or since 1859, the price of foreign steel of similar qualities has been very considerably reduced through the effect of the American competition, and that the whole country in this way has gained more than sufficient to counterbalance the tax levied as a protection for the American steel manufacture, which has grown up under its influence.

At the time Mr. Wells gave these deliberate utterances to the world, as the result of his official investigations, he had not sent himself to England at the public expense, and there received a round of flattering entertainments that described the circle of British Free Trade propagandism. When he had studied the needs of our industrial condition from a patriotic standpoint, he was convinced that the Protective policy had been a gain to the whole country; but when he studied the same subject from the British point of observation, he discovered that the same policy had been a loss and a snare to his countrymen. At a meeting of the American Iron and Steel Association, held in Washington, January 16, 1867, Mr. Wells had said:

I desire here and now, unequivocally and unreservedly, to declare that, in the British sense of the word, there is no Free Trade in me. From my earliest childhood I have been taught the value of the doctrines of Protection, and it has been my fortune to sit at the feet of that great teacher of political economy, Henry C. Carey, and learn from him the great principles on which these doctrines are founded—the complete and universal harmony between all the producing interests of the country. In that faith I am as strong to-day as I ever was.

But it required only a visit to England, and audience given to the Manchester school of political economists, to wipe all these convictions from his mind as a sum sponged from a slate. He came back with British lenses before his eyes, and never since has been able to see any American manufacturer except through their distorting medium. Yet the original judgment of Mr. Wells is fully supported and vindicated by the present state of the facts. He declared in 1867 that the welfare of the higher grades or qualities of steel was involved in a maintenance of highly Protective duties. Now, that very class of steel is now quoted in New York at the following prices: American, 14½ to 15½ cents per pound in currency; English, 17¾ cents gold. Here is another example of the cheapening effects of Protection to home industry. Of course, when steel is so reduced in price, iron must be still lower; hence we are able to say that iron rails are now selling at $45, currency, per ton; and that no foreign rails can be imported to compete with these figures. Our Protective system, in less than fourteen years, has made us independent of European sources for our supply of railroad iron, and equally independent as regards steel rails. To-day we can build railroads cheaper with home-made steel rails than some years ago we could with imported iron rails. How, then, can road-bed, rails and rolling stock be so swollen in cost by a high tariff as to make cheap transportation an impossible thing?

Increased cost in production, it may be worth while to relate, is not, as Free Traders dogmatically assert, an inseparable adjunct of our Protective system. The lowest average price at which foreign rails, of iron, were ever before now sold in the American market was $44 per ton, during the first seven months of 1861. That was in gold. If, as is the fashion, we reduce the present currency price of American rails, or $45 per ton, to equivalent coin, at the prevailing quotation of 114 for gold, we shall have $39.46 as the result, or $4.54 per ton lower than the lowest price of foreign rails ever before reached. Yet we are told, with an immense expenditure of emphasis, that our Protective system is a device to plunder the masses of the people, and especially Western farmers, in the spurious name of industrial development; and that the tariff duties on the materials which enter into the construction of railroads are so extravagantly high that freights and fares are forced to be raised in proportion. For all that, the proofs of the allegation are as hard to find as the traditional needle in the traditional haystack.

It needs to be noticed that while cost in production has receded in the United States it has advanced in Great Britain—that central seat of Free Trade propagandism, plainly showing that the system of tariff adopted there does not necessarily promote cheapness, nor prove a safeguard against a considerable rise in prices. For many years her cheap labor operated with all the force and effect of a Protective tariff. Now that her labor is becoming dear as compared with that of various European countries, she encounters many of the difficulties that would beset the United States under her system of duties. Never before has her manufacturing ascendancy been in such mortal peril as it is to-day; and bankruptcy of great iron establishments is assuming an alarming extent and frequency.

The following statement from the editorial columns of the Philadelphia *North American*, May 28, 1874, illustrates the independence achieved by this country in the manufacture of Bessemer steel rails.

Large orders for steel rails have been taken by the Chicago and Joliet rolling mills, amounting to seventy-five thousand tons, at forty per cent. less than the ruling rate of 1873, and twenty-five per cent. less than the current rate for foreign rails. They are for the Rock Island, Alton, Illinois Central, Michigan Central, and the Central Pacific, and Union Pacific, and are to replace old iron rails.

On inquiry at the proper quarters, we find the above statement to be correct, with the exception that the Chicago and Northwestern Road should have been included in the list. These facts speak trumpet-tongued of the cheapening tendencies and effects of Protective duties. Transportation is in process of being cheapened by our tariff system, which has fostered the manufacture of steel rails upon American soil, multiplied rolling mills, led to important improvements in mechanical appliances, energized home competition, brought down prices, and given the domestic market to the domestic producer. If in all these years we had had the kind of tariff which Free Traders extol, our railroads would now be at the mercy of foreign high prices, and might be paying the English for steel rails $120, gold, per ton, delivered in New York as one Western road actually is, in fulfillment of an old contract.

CHAPTER XX.

SELL DEAR; BUY CHEAP.

IN another column will be found a letter from a correspondent signing himself "Traveler," who, after according to this paper honesty of purpose and ability in argument, joins issue with the positions occupied by the *Inter-Ocean* on the currency and the tariff. Before taking up the point which we intend chiefly to notice, we shall answer his fundamental proposition, thus stated:

Government exists for the protection of individuals. This is the sum and substance of its object, and beyond this it should not go.

Such, however, is not the language of the preamble to the organic law of the Union, in these words:

We the people of the United States, in order to form a more perfect union, establish justice, insure domestic tranquility, provide for the common defense, promote the general welfare, and secure the blessings of liberty to ourselves and our posterity, do ordain and establish this Constitution for the United States of America.

It thus appears that *our* Government, at least, looked, in its institutions, not to *individuals*, but to *communities*—that it exists to protect and preserve *society;* that its seminal design was to create a *nation* out of fragmentary bodies of people; to that end possessing and maintaining relations which are external or foreign, besides those which are internal or domestic; and sometimes, to the same end, depriving individuals of life, or shutting them up in prison, condemned to hard labor for a term of years. All this embraces much more than is embraced in the statement, that "government exists for the protection of individuals." Government exists for many things more, among which are various public interests, as harbors, light-houses, military roads, the national lands, education, and the like, these extending far beyond the mere "protection of individuals," and requiring the exercise of additional functions. The revenue system and the currency system that may be adopted by Government must be judged, therefore, by their effects upon the *mass* of individuals—upon the *general* welfare—upon the *body politic*. Our objection to the Free Trade policy is, that, wherever it has been even partially practiced, it has operated as a retrogressive force, as is exemplified in the cases of Turkey, Portugal, and Ireland, and is beginning to be manifested in the case of England herself. Our correspondent says further:

> The farmer, having raised his produce, should first sell it for all he can get in cash, and then spend his surplus profits hiring or purchasing *for cash* where he can do so the cheapest.

Here we have a repetition of one of the stereotyped errors of the Manchester school of political economists. To buy in the cheapest market, and to sell in the dearest, is a maxim of the first rank in the Free Trade philosophy. This maxim, however high-sounding and seductive it is in theory, violates in practice its own principles and precepts. Those who buy cheap must necessarily purchase of those who sell cheap, and who, therefore, do not sell in the dearest market, according to the specific injunction laid upon them as part of their duty to themselves. On the other hand, those who sell dear must find purchasers who do not buy cheap, and who consequently set at naught the exhortation to buy in the cheapest market. By the terms of the maxim, all buyers are to buy cheap, and all sellers are to sell dear. How all members of a community are to buy in the cheapest market while all are selling in the dearest, or how all are to sell in the dearest market while all are buying in the cheapest, is one of those puzzles which it would require a miracle to solve. Yet this impracticability is exactly what Free Trade philosophy demands shall be done every day, and what it extols as the very quintessence of commercial wisdom.

When we attempt to apply the maxim to different countries or communities, the same inconsistencies appear. If England should buy cheap of the United States, and sell dear to the United States; or should buy of us dear and sell to us cheap; or should buy of us dear and sell to us dear; or should buy of us cheap and sell to us cheap, either England or the United States would, in each case, violate the teachings of the Free-Trade maxim. If England should buy cheap of the United States, and not sell to us at all, but sell dear to some other country, then both the United States and that other country would set at naught the exhortations of the maxim, the United States failing to obey the injunction to sell dear, and the other country the injunction to buy cheap. Indeed, no country or countries can possibly buy cheap and sell dear, unless some other country adopts into practice a course at war with the teachings of the maxim; for no country can *buy cheap* or *sell dear* without finding some other country that will reverse the procedure by *selling cheap* or *buying dear*—that is to say, by doing the exact contrary of what the Free Trade maxim enjoins upon it as a duty. To be an operative maxim, not contradicting in practice its own precepts and obligations, it must necessarily be restricted to a few countries, or to all countries except one, the theory in such case being that certain countries are endowed with special, exclusive rights of trade, and there being a limitation upon the privileged countries that they can not carry on commerce with one another, else there would be an immediate violation of the rule; for, the moment any one of the number *bought cheap*, another would have to *sell cheap*, which certainly would not be *selling dear;* and, on the other hand, whenever one of these countries *sold dear*, some other of the countries would have to *buy dear*, which assuredly would not be *buying cheap*, according to the specific exhortation of the maxim. Now, as it is impossible for all countries together and at once to be buying in the cheapest market and selling in the dearest, there is no way to make the rule operative at all, unless it shall be confined to a certain number of countries excluded from commercial exchange with one another, yet clothed with exceptional privileges of trade with some other country or countries. A, B, and C—separate and distinct countries—might be able to buy cheap of and sell dear to D, E, F, G, and H, but these eight countries could not possibly do so among themselves indiscriminately. Then, as the Free Trade maxim can not be made of universal application, will somebody—our correspondent, for instance—be so good as to indicate what countries are to be privileged to buy cheap and sell dear, and what countries—shall we say the United States, for one?—are submissively to sell cheap and buy dear; and on what principle of morality he justifies such limitations—*necessary limitations*—of the operation of his pet maxim? Or, failing in this, will he, or any other Free Trader who supposes he understands the process, be so obliging as to explain how it is practicable for all commercial nations to comply with his maxim of trade?—how all nations are to buy in the cheapest market while all are selling in the dearest, or how all are to sell in the dearest while all are buying in the cheapest?

It thus appears that this Free Trade maxim has neither utility nor sense, when universally applied. It is merely a rule of selfishness and aggrandizement; for, to buy cheap is to buy at less than the market or real value; and to sell dear is to sell at more than the thing is worth. One or several countries may thrive by practicing such sharp bargain-making, but it must be by victimizing other countries, either by buying of them cheap, or by selling to them dear. Both parties to commerce so carried on could not gain. All the advantage would be on one side; all the deprivation on the other. The Spaniards, discovering America, practiced on this maxim in exchanging a few glass beads and some gaudy trinkets for commodities of great commercial value. Really, the precept is the essence of self-seeking, constitutes the very spirit of making haste to be rich, and could not have originated in any feeling of broad philanthropy, or of doing unto others as you would have others do unto you.

Nor is this all. The maxim contradicts another Free Trade maxim of equal authority, and of equally binding obligation. Those who agree that all persons should buy in the cheapest market and sell in the dearest likewise insist that consumers are more numerous than producers, and that the rights and interests of the former are paramount to those of the latter, so that the need of consumers to have cheapness is to be consulted before the need of producers to have dearness. Consequently, whenever a commodity is sold dear, the right and interest of the consumer to get it cheap are invaded, and he has just cause of complaint against the seller. Notwithstanding this, a fundamental maxim of Free Trade lays upon the seller an imperative injunction to sell dear, and also upon the purchaser, at the same time, to buy cheap. If producer or dealer sells any vendible cheap, he violates a duty he owes to himself; if either sells such vendible dear, he violates a right belonging to consumers, and which he, as an integral unit of society, owes to them. If the consumer buys cheap, he induces the vendor to break the precept that exhorts him to sell dear; or, should the consumer buy dear, then he upholds a wrong to his class which society should see put down. Whether the dealer sells either dear or cheap, and whether the consumer buys either cheap or dear, they both violate each other's rights or privileges, considered from a Free Trade stand-point. In fact, there can be neither buying nor selling without violating a precept which the Free Traders esteem as of the highest importance. Thus does visionary theory butt its head very hard against practice, and experience, and common sense, getting badly worsted in the encounter. Such is the result of closet-logic, inventing maxims of conduct without consulting experience.

The truth is that the price of a sale always embodies a compromise or adjustment of conflicting interests between buyer and seller, unless some element of compulsion enters into the contract—a compromise which varies according to the knowledge, needs, and other surroundings of the parties to the transaction. It is one of the crowning merits of the Protective policy that it more and more tends to establish circumstances which result in increasing the prices of raw materials and the rate of wages, and in reducing the prices of finished products. An example of these consequences is found in the case of the woolen manufacturer in Indiana, who, in 1860, under partial Free Trade, paid 25 cents per pound for wool and $1.50 per day for labor, wholesaling 9-oz. jeans at 60 cents per yard; yet who, in 1874, under a Protective tariff, paid 50 cents per pound for wool and $3 per day for labor, wholesaling 9-oz. jeans at 50 cents per yard. A whole bookful of impracticable maxims weighs as nothing in the scale against this single fact.

CHAPTER XXI.

AMERICAN POTTERY.

BEFORE the series of Protective tariffs which began their course in 1861, this country was disgracefully dependent upon foreign manufacturers for its supply of earthen, stone, and porcelain ware. Such was the state of facts, notwithstanding that the United States hold the greater part of the available coal of the world, near to inexhaustible quantities of varied clays and other potting materials, besides possessing unusual attractions for skilled and ordinary labor. Under Protection, the whole face of the situation has been changed. Says William P. Blake, in his general survey of the ceramic arts, which prefaces his report on that department of the late Exposition at Vienna:

The industry, especially in the direction of earthenware, and the common cheap pottery, such as Rockingham, yellow-ware, and stone-ware, has increased rapidly of late years, under the stimulus afforded by the tariff and the premium on gold. According to the last census, there were 777 establishments for the manufacture of stone and earthenware distributed through the several States, the highest numbers being 170 in Ohio, and 198 in Pennsylvania. Only fifteen are reported in Massachusetts. Eighty-two steam-engines, with an aggregate of 1,586-horse power, were in use, besides eight water-wheels of 122-horse power. Hands employed, 6,116; capital invested, $5,294,398; amount paid in wages, $2,247,-173; materials are valued at $1,702,705; value of the products, $6,045,536. The number of persons reporting their occupation as potters is 5,060.

The following are the chief points at which the potteries are located: in New Jersey, at Trenton, Jersey City, and Gloucester; in Ohio, at East Liverpool and Cincinnati; New York, in the city, and at Flushing and Greenpoint, L. I.; Pennsylvania, Philadelphia and Pittsburgh; Illinois, Peoria; Maryland, Baltimore; Massachusetts, Boston; and in Missouri at St. Louis. In 1872 it was estimated that there were 148 kilns in seven States, capable of producing at the rate of $30,000 annually per kiln, which would amount to $4,440,000 per annum, and would use 75,000 tons of coal, and 75,000 tons of clay and other materials.

The industry has taken root firmly in New Jersey, at Trenton, and bids fair to thrive permanently. That locality offers the advantages of extensive deposits of the finest clays; cheap transportation by water, as well as by rail; and the proximity to the coal region and to two large cities combine to foster its growth, and to make the locality the Staffordshire of the United States.

Meantime, with all this growth of domestic production, the home market is expanding faster than the rate of supply from home sources, as is evidenced by the values of earthen, stone, and china ware imported into this country during the following fiscal years: In 1869, $4,372,607; in 1870, $4,388,771; in 1871, $4,681,376; in 1872, $5,270,785; in 1873, $6,015,945, and in 1874, $4,882,-355: aggregate for six years, $29,611,839. Of this very large aggregate, Great Britain contributed to the amount of $23,766,-646, distributed annually as follows: in 1869, $3,687,532; in 1870, $3,571,859; in 1871, $3,889,397; in 1872, $4,151,150; in 1873, $4,646,688, and in 1874, $3,820,020. But Englishmen are not satisfied with this close approach to a monopoly of the patronage we bestow upon foreigners. On the contrary, their souls are aglow with indignation at the restrictions placed on international commerce by our iniquitous system of tariff. So deeply are these cosmopolitan philanthropists imbued with feelings of *disinterested solicitude for our industrial development and welfare*—to be brought about by Free Trade!—that several English manufacturers last year visited the United States to familiarize themselves with the nature and extent of the American competition which had begun to overmaster them in our markets, and which was threatening to annihilate their business in this country. These large-hearted, generous-minded, sympathetic Britons were so overwhelmed with a sense of the folly, delusion, and injustice embodied in our Protective policy, that they returned home with the noble resolve, as they expressed it, "to unroof the potteries in Trenton, and destroy the plant of capital there." This reveals the genuine spirit of British Free Trade, whose occult meaning is that we should send our clay, our sand and our coal over the ocean to be worked into objects for our daily use, rather than manufacture for ourselves. When we shall have reached that degree of enlightened sagacity, of circuitous exchange, and of practical common sense, we may hope to be kindly granted the privilege of exporting our wheat to England, there to be converted into flour, and thence to be brought back, before we convert it into bread. Then we shall be prosperous, indeed! Such are the blessings of Free Trade. Let us learn, therefore, that, the more commodities are moved to and fro, the lower will be the prices, and the more extensive the commerce :—in other phrase, take the manufacturer away from the side of the consumer; interpose three or four thousand miles between the two; increase the number of middlemen and the demand for the machinery of transportation: at once you will set in motion a series of causes that tend inevitably to cheapness. That is the milk in the Free Trade cocoanut. In its last analysis, British Free Trade means that the United Kingdom—with its vast mercantile navy, its numerous insurance companies, its extensive network of branch houses, agents, factors, and banking facilities, its astute devices of consular action and of diplomatic manipulation, and its prodigious resources for manufacturing—shall become a sort of commercial sponge to soak up the profits of the world's exchanges. The threat, "to unroof the potteries in Trenton, and destroy the plant of capital there," represents the spirit of that so-called Free Trade which would crush out the reproductive arts in every rival country, merely in order to give Great Britain more markets in which to sell her finished products.

A gentleman in Philadelphia—Horace J. Smith, Esq.—who has made a special study of the pottery industry in this country, sends us the following statement:

Crockery manufacturers have already so cheapened their processes, under the stimulus of a home competition, as not only to drive English goods of the lowest grades entirely out of the market, (viz: yellow and Rockingham,) but they have almost entirely shut out the next better grade (the C. C. or cream-colored) from importation. They are successfully competing with the best English manufacturers for the American demand for white stone-ware, as well in price as in quality; so that, with the maintenance of Protection for one decade longer, we may hope to attain complete commercial independence of England in this respect. The manufacture of porcelain, too, which is the finest product of the ceramic art, has already taken root in the United States; and in the near future we may expect that neither ware from "China," nor "Delft" from Holland, nor even "Queensware" from monarchical England, will find a place in American homes.

Closer relation between the consumer and the crockery manufacturer has so forced itself upon the latter as a necessity for his prosperity, that it is in contemplation to establish potteries either in Chicago or Milwaukee, or in both. As an additional inducement, the ground for a factory has been offered as a gratuity to an enterprising Eastern manufacturer, provided he will establish himself at Milwaukee. Whichever point is selected—whether one of the two named, or some other offering greater inducements of location or of pecuniary assistance—may anticipate a promotion of its general interests in various and unexpected directions. A permanent sign in one of the main streets of Trenton points the way to one of its many potteries, with this additional notice to the farmer—"Oat straw bought." So apparently insignificant an item (yet straws show which way the wind blows) points the farmer which way his interest lies. Without a diversification of industries, the straw may rot, (or even, as has been the case, be burned to get rid of it,) the clay lie inert and valueless in the bank, the building materials never be assembled to form cosy homes, no coal be mined, no railroads built, and no enhanced value put upon the farmer's acres. With the pottery established, comes also a demand from its operatives for every varied product of the farm: "walk-away crops," such as cattle, are needed and kept at home; the full-uddered cow lows for the milkman; and they do say that even water to thin down her lacteal richness comes into demand!

As has been proved by J. W. Foster, Esq., the "pre-historic races of mound-builders" maintained a dense population in the Mississippi Valley; but we can never reproduce this condition if we separate the producer from the consumer by long distances, as is enforced by remaining a purely agricultural people. By all means, then, invite the potter, the tanner, the shoemaker, the spinner, the clothier, the hatter, the sugar-refiner, the butcher, the baker, the candle-stick maker—in short, all classes of manufacturers—to come and make their abode close by the Western farmer.

As yet, we have made scarcely more than a beginning of the pottery manufacture in this country, which is peculiarly rich in all the elements requisite for the successful and permanent development of the ceramic arts. Even as early as 1770 it became evident to the British potters that the pottery industry might be started in America to the detriment of their trade; and Wedgewood, whose name is identified with the growth of artistic pottery in Britain, wrote as follows:

> The trade to our colonies we are apprehensive of losing in a few years, as they have set on foot some pot-works there already, and have at this time an agent amongst us hiring a number of our hands for establishing new pot-works in South Carolina. They have every material there, equal if not superior to our own, for carrying on that manufacture. We can not help apprehending such consequences from these emigrations as make us very uneasy for our trade and prosperity.

In Great Britain the ramifications of the manufacture are quite numerous, embracing, as they do, thirty-nine distinct trades engaged in the potter's and connected arts, besides seventeen other occupations employed in supplying raw materials or machinery. With due Protection we may expect to reach a still higher degree of development, and ultimately to become extensive exporters, not only of crockery, but also of those porcelain wares which embody artistic excellence, and sometimes the triumphs of genius for fictile design. So far every advancing step in the industry on our soil has been rapidly upward. Already the inventive talent of our countrymen has very largely substituted labor-saving machinery for the toil of human hands so prevalent in Europe, with the immediate effect of greatly cheapening cost in production, and thus placing American manufacturers more nearly upon an equality of competition with the long-established factories in the Old World, based on low wages. In the production of several classes of wares an overwhelming advantage has been gained in this way. An instance of this appears in the fact that porcelain door-knobs have fallen in price at the pottery works from $12 to $3 per thousand. Other hardware trimmings are now sold for less than one-quarter of the price they brought from 1845 to 1848, when the duty was much lower than it now is. With growing experience in business, and with improved methods in manufacture, results of such kind will be multiplied, until expansion of demand and domestic rivalry shall so permanently and comprehensively achieve cheapness as to shut out foreign competition, besides giving us a large export trade. Then we shall not need to go to Minton, Hollins & Co., of Great Britain, in order to procure tile pavement for our public buildings, as we had to do for the new capitol at Washington. But without Protection to home industry, we would have to buy from England not only tiles, but even the material for our national flag, as used to be the shameful case.

CHAPTER XXII.

A PROTECTIVE TARIFF AND EXPORTS.

GRANT, NEB., *December* 4, 1874.

To the Editor of the *Inter-Ocean.*

If manufacturers in this country require such a high Protective tariff, how can they afford to send their goods to the Eastern Continent, and there compete with foreign manufacturers? Please answer in the *Weekly Inter-Ocean*. P. FORD.

AS yet, a very small fraction of our finished products is exported to the Eastern Continent. We have made a beginning in that way, and it is increasing; but the great bulk of our exports of manufactures goes to foreign countries on the Western Continent, and to the adjacent islands. For example, we exported, during the fiscal year 1874 various passenger and freight cars for railroads to the number of 1,083. Of these, 6 went to Belgium, 10 to Germany, 12 to England, and 1 to Scotland, or only 29 in all to Europe. The rest were distributed, 81 to the Argentine Republic, 31 to Brazil, 18 to Central American States, 286 to Chili, 222 to Nova Scotia and New Brunswick, 188 to Quebec and Ontario, 32 to Mexico, 68 to Peru, 67 to Cuba, 10 to the United States of Colombia, and 51 to Uruguay, or 1,054, being more than 97 per cent. of all, to ports in North and South America and Cuba. However, we send large quantities of clocks, sewing machines, agricultural implements, and some other articles to the Eastern Continent.

A comparatively small part of our exports takes the form of fabrics. During the fiscal year 1874 we exported domestic products to the enormous value of $693,039,054. Of this vast sum $3,310,388 represented living animals, $161,198,864 breadstuffs, $211,223,580 raw cotton, $41,103,516 petroleum of various grades, $78,328,990 provisions, $2,758,933 spirits of turpentine, $8,135,320 tallow, and $32,968,528 tobacco and its manufactures, making a total of $539,028,119, the whole being either raw materials or manufactures but little advanced beyond raw materials. Even of the remaining $154,010,935 of exports, a considerable portion stands for like products, as $3,285,210 for naval stores, and $21,353,721 for timber, lumber, masts, spars, staves, and other forms of wood, only $1,882,767 of the sum representing furniture, and $1,772,410 other manufactures. These facts show that our manufacturing industry, notwithstanding its great development in recent years, is still merely in its infancy, and needs the fostering encouragement of Protective legislation.

The United States are extensive exporters of food and of raw materials. These are mainly taken by the great manufacturing nations of Europe, which are extensive exporters of finished products. Much of our foreign trade is imbued with the policy of "selling a rabbit's skin for a sixpence, and buying back the tail for a shilling." This false mode of exchanges can be superseded only by the multiplication of the arts and the sciences upon our own soil. To accomplish such result, the constant aim should be to bring the loom and the anvil everywhere into close neighborhood to the plow and the harrow; for then alone can we consume our own food and reproduce our own raw materials, exporting the surplus in its highest forms. Then alone can labor find steady employment in its various departments, all aptitudes be set at work, and the productive forces kept fully occupied; for laboring power, of whatever kind, like time, unless utilized at the very moment of existence, drifts unproductively into the past, and is lost beyond recall, the country and the man both being poorer by what might have been produced by the use of the power. It is the province of a high Protective tariff to aid in creating and in maintaining those conditions which result in regular employment for all, and the highest measure of production and consumption, whose other name is general prosperity. These preliminary observations are essential to a thorough understanding of the subject.

Two sets of agencies, co-operative in this country, under a system of Protection to home industry, enable our manufacturers to export their finished products, thus overmastering the competition of cheap labor in foreign countries. These agencies are:

(1) Superiority of make, as regards both material and workmanship, and complete adaptability to intended uses, causing the

article to be more desirable than its European rivals, even though they be offered at a lower money price.

(2) Labor-saving machines, which may be tended by one or more persons, sometimes even by a child, yet which will do the work of five, ten, fifteen, twenty-five, or a hundred able-bodied men in a day, reduce cost in production to a small fraction of its former amount, and overmaster the competition of manual processes still in vogue in Europe.

Our common chopping axe is an example under the first proposition. An immense amount of intelligence has been expended upon this instrument of toil. Its weight and shape, the bevels of its sides, the location in the general mass of the hole which receives the helve, the position, length, and curvatures of the handle, have all been designed to accomplish the most work with the least outlay of strength. These adaptabilities have been largely neglected in the manufacture of the foreign axe. Hence it is that the Sheffield (England) *Telegraph* said, not many weeks ago: "The steel of an American axe is so superior to that of an imported axe, that no pioneer who understands his business will ever carry any other with him into the wilds." And hence it is that the London (England) *Times* said editorially, in the early part of 1871:

The Americans succeed in supplanting us by novelty of construction and excellency of make. *They do not attempt to undersell us in the mere matter of price.* Our goods may still be the cheapest, but they are no longer the best, and in a country where an *axe*, for instance, is an indispensable instrument, *the best article is the cheapest, whatever it may cost.* Settlers and emigrants soon find this out, and they *have* found it out to the prejudice of Birmingham trade.

An example of the advantages derived by this country under our second proposition is to be found in the manufacture of clocks, in which we outrival the world. All the important parts of these cheap time-keepers are made by labor-saving machinery. In 1841, Connecticut clocks were just beginning to be exported to England, where they sold at first at an advance of a thousand per cent. on cost.

This export trade is constantly increasing in quantity and in the number of foreign markets. During the four years ended June 30, 1874, we exported clocks and parts thereof to the value of $3,-107,712, fractionally distributed as follows: In 1871, $552,155;

in 1872, $679,162; in 1873, $868,888; in 1874, $1,007,507. In the last year England took to the value of $533,600; Scotland, $34,205; Germany, $103,688; China, $12,461; Japan, $61,485.

Another striking example under our second proposition is to be found in the rising manufacture of crockery in this country. We quote from an editorial article in the Philadelphia *Press*, Nov. 16, 1874:

> Another industry, too, may wane in England, to grow correspondingly great in the United States, and this is the crockery manufacture. It is well named *manu*-facture, as so much is done by hand, and so little (at least in the old country) by machinery. At Trenton, N. J., some unprogressive European potters were settled, as well as some ingenious Yankees. It is, or at least was, a while back, amusing to walk from the factory of one to the factory of the other. The original methods in use, perhaps from the infancy of our race, were pursued in the one, even as they are to-day at Staffordshire by such distinguished potters as Minton, the maker of encaustic tiles for the world. One of the first processes in making pottery is to thoroughly intermix the clay and water, and this was done by some of the Trenton potters, as it is still done in England, by men stirring the ingredients together with paddles in large vats. When the clay and water have come to the consistency of cream, the men ladle the "slip," as it is called, from the vat into the sieves. Then shaking the sieve, which consists of fine cambric, over another vat, after the fine material has passed through, the coarser particles are emptied out, and the sieve refilled. Such has been the process for ages, and conservative Englishmen pursue it to this day. But stepping, as we say, from one pot-house to another, we find steam rotating paddles in the slip-vat, and steam pumping the slip up into sieves, which are themselves swiftly agitated by steam. Such are the contrasts exhibited between European and American mechanics and manufacturers.

Now, a high Protective tariff promotes and maintains the conditions amid which such industrial developments can take place. No one will begin a manufacture without reasonable prospect of finding a market for his product. If foreigners occupy and engross his domestic market, he can not expect, when he may be undersold at home, to secure a profitable market abroad, in competition with those very foreigners. Under such circumstances he will make no venture. Should he do so, he will run the risk of being ruined by the intolerance of his foreign rivals, who will combine to prevent the success of his infant enterprise, by overwhelming him with exceptionally cheap prices, until he is compelled to retire from the unequal contest, whereupon those rivals, rid of the competition, resume a monopoly control of our market, raise their prices, and reimburse themselves from consumers for their losses in breaking down the American industry. Such is the certain result when there are no customs duties. But if a high Protective tariff should be laid on the import of such articles as we can ourselves produce, then the crushing-out process becomes very expensive, and is almost sure to fail. Consequently, capital is encouraged to invest in manufacturing undertakings; domestic labor receives employment; home resources are developed; industry is rapidly diversified; purchasing power is created by the payment of wages; production and consumption grow apace; commerce becomes active and widespread; inventive genius is everywhere stimulated; labor-saving machinery is multiplied in all directions; finished products grow cheaper and cheaper; the sciences and arts flourish; prosperity is universal; and the freedom of man expands and intensifies. If we have not had this full measure of blessing, it is because our tariff policy has been capricious, vacillating and unreliable, thus creating fluctuations of effect, and leading to conflict and disorder.

Manufacturers, considered as a class, require a high Protective tariff as a means of securing a market for their products. If a man makes 100 tin pans a week, which he must sell at an average profit of twenty-five cents, in order to carry on his business and live, he would be far better off if he could make and sell 10,000 pans a week at a profit of one cent each; for he would gain, in the former case, only $25; in the latter, $100. Not only would he benefit his customers; he would also give increased employment and wages to labor. Every additional mechanic he would employ would require additional food, clothing, etc., to be supplied by somebody else. By such interaction and reaction, all persons willing to labor may ultimately find steady employment and good pay. Then each produces something to be exchanged for something else. The greater the number of commodities produced, the greater, other things being equal, will be the number of exchanges. Commerce tends, therefore, to grow with the increase of production; and production tends to increase under a high Protective tariff.

Under a non-Protective tariff the result is different. Then we surrender to European countries the bulk of our manufacturing industry. We hire cheap foreign labor to convert raw materials into finished products for our use, instead of bestowing the employment upon labor at home; doom a multitude of our people to partial idleness; reduce the general rate of wages; cripple the productive forces; and compel a shrinkage in consumption. Therefore our manufacturers require a high Protective tariff, because it will give them possession of their home market; our laboring classes require it, because it will secure for them full employment, with good pay; and our whole people require it, because it is conducive to substantial and permanent prosperity.

CHAPTER XXIII.

WHY OUR MANUFACTURERS WANT PROTECTION.

THE following query embodies a difficulty which has puzzled many who are unfamiliar with the comprehensive workings of our tariff system:

CHICAGO, *June* 28, 1875.

To the Editor of the *Inter-Ocean*.

You say that high Protective duties operate to cheapen the prices of manufactures. If so, why do our manufacturers want such duties? W. B. TERRY.

Our manufacturers want Protection, although it operates to reduce prices, because the practical benefits of Protective duties are found, not so much in diminishing regular importations made by our own merchants, with a full knowledge of the wants of the country, but in securing our ports of entry from being flooded with surpluses *forced* upon us by foreigners, to relieve their own superabundance and sustain their home prices, or to overwhelm our industries at any temporary sacrifice, that they may command our markets. England, our principal competitor, from whom we have derived, within the last twenty-four fiscal years, 39.391 per cent. of the total value of our imports, can produce a long list of fabrics cheaper than we, on account of her far cheaper labor, much lower interest on money, and immense accumulation of capital, and can undersell our manufacturers on our own soil *when she chooses*. It invariably happens that her lords of the loom and her earls of the rolling-mill choose to do so whenever the rising prosperity of our manufacturing enterprises begins to supply the needs of American consumers. Then a vast aggregate of foreign commodities is ruthlessly thrown upon our well-stocked home markets, at lower and still lower prices, thus ruinously forcing down to that level the selling values of our whole domestic product, even although it may be tenfold the quantity of the imported articles. This cheapening process is kept up until our home producers are worn out with the unavailing struggle, and are driven out of business or into bankruptcy. Meantime, the British schemers have themselves lost heavily, but can well endure the strain of the losses, on account of their much larger capital, coupled with profits on their trade carried on with other countries, producers in the United States generally not possessing such a compensating advantage. The victory thus won, leaving our manufacturers crippled or prostrated, their foreign rivals boldly step in for our trade, which they monopolize with unsparing rapacity, prices often being advanced, under these circumstances, to 50 and even 100 per cent. on what they had been before the struggle began. The temporary cheapness which beguiled and deluded our consumers disappears altogether, instead of which they find themselves the helpless dupes and victims of a heartless foreign extortion. With possession and control of our market assured, British producers reimburse themselves for their losses out of the pockets of our people, who are thus made to pay the cost of breaking down the barrier erected to shield them against the merciless practices of transatlantic rapacity. Says B. F. French, in his "History of the Iron Trade of the United States:"

> It is an established fact that, in many departments of English industry, those who are interested will carry them on at a loss for years, to aid in retaining markets from which they are in danger of being excluded by commercial restrictions or industrial competition; and scarcely a branch of industry has sprung up in the United States which has not, at first, had to encounter a severe struggle in consequence of the foreign article being reduced in price below that with which it had been expected to compete.

That this statement, in regard to certain manufacturing capitalists in England, misrepresents neither their acts nor their motives, is made evident by the following open avowal which we find in an official document. It occurs in a report presented to Parliament by a commission appointed in 1854 to inquire into the operation of an act relating to the mining population. Here it is:

> The laboring classes generally, in the manufacturing districts of this country, and especially in the iron and coal districts, are very little aware of the extent to which they are often indebted for their being employed at all to the immense LOSSES which their employers voluntarily incur in bad times in order TO DESTROY FOREIGN COMPETITION, AND TO GAIN AND KEEP POSSESSION OF FOREIGN MARKETS. Authentic instances are well known of employers having, in such times, carried on their works at a loss amounting

in the aggregate to three or four hundred thousand pounds sterling in the course of three or four years. If the efforts of those who encourage the combinations to restrict the amount of labor and to produce strikes were to be successful for any length of time, the great accumulations of capital could no longer be made *which enable a few of the most wealthy capitalists to overwhelm all foreign competition in times of great depression*, and thus clear the way for *the whole trade* to step in WHEN PRICES REVIVE, and to carry on a great business before *foreign capital can again accumulate* to such an extent as to be able to establish a competition in prices with any chance of success. *The large capitals of this country are the great instruments of warfare against the competing capital of foreign countries*, and *the most essential instruments now remaining* BY WHICH OUR MANUFACTURING SUPREMACY CAN BE MAINTAINED; the other elements—cheap labor, abundance of raw materials, means of communication, and skilled labor—being rapidly in process of being equalized.

Under our present tariff system, these schemes of spoliation are impracticable, so far as the United States are concerned. The duties which must be paid at our custom houses, before foreign goods can gain admission into this country, make a crusade upon our industries, in the form of a ruinously cheapening process, too costly to be undertaken, and too unlikely of success. Instead of that, we have many and fulsome praises of the assumed blessings of Free Trade. Unrestricted competition with pauper labor in Europe, and with low interest on money, and with immense accumulations of capital, is to pour a flood of cheap commodities into the United States, to the great advantage of *consumers*. But what, meantime, is to become of American *producers*? When they shall be ruined or driven out of business by the deluge of temporary cheapness, what is to prevent the British manufacturers from making a ruthless monopoly out of their possession and control of our market, prices being forced up as high as the power of resistancer will endure? Under such circumstances, the cost of what is miscalled Free Trade—what is really the slavery of commerce—would far exceed the cost of maintaining and prospering our own industries. The fact is, the people most to be benefited by a repeal of the duties on iron, steel, cottons, woolens, and other staple articles, are the monopolizing capitalists across the ocean. Intelligent Englishmen fully comprehend this truth. What they think, in the sincerity of their hearts, of the policy of Free Trade, is manifest from the following extract from a speech of a member of the British Parliament, delivered at a time when the United States had adopted the policy of Protection, and quoted by Henry Clay, in 1832, in the Senate of the United States.

It was idle for us to endeavor to persuade other nations to join with us in adopting the principles of what was called "Free Trade." Other nations knew, as well as the noble lord opposite, and those who acted with him, what we meant by "Free Trade" was nothing more nor less than, by means of the great advantages we enjoyed, to get a monopoly of all their markets for our manufactures, and to prevent them, one and all, from ever becoming manufacturing nations. When the system of reciprocity and Free Trade had been proposed to a French ambassador, his remark was, that the plan was excellent in theory, but, to make it fair in practice it would be necessary to defer the attempt to put it in execution for half a century, until France should be on the same footing with Great Britain, in marine, in manufactures, in capital, and the many other peculiar advantages which it now enjoyed. The policy France acted on was that of encouraging its *native* manufactures, and it was a *wise* policy; because, if it were freely to admit our manufactures, it would speedily be reduced to the rank of an *agricultural nation*, and therefore, a *poor nation*, as all must be that depend *exclusively upon agriculture.* America acted, too, upon the same principle with France. America legislated for futurity—legislated for an increasing population. America, too, was prospering under this system.

Here we find plenty of reasons why our manufacturers want Protection, even though it operates to cheapen prices. They are relieved thereby from foreign fluctuations and excesses, from foreign machinations and interference. No matter how capable or how energetic any people may naturally be, nor how favored in position, or climate, or soil, their industrial capacities can never be fully developed under a vacillating and uncertain public policy. Men do not embark either capital or skill in enterprises liable at any time to be defeated by inconsiderate or unfriendly legislation. A stable order of things and a well-founded confidence in the future are all-essential conditions of manufacturing success. Such stability and such confidence are supplied by sufficiently Protective legislation. Under it, prices have always, in general, exhibited a downward tendency as regards all finished products, while farmers have received better prices for their produce. So long as the Protective duties remain upon the national statute-book, there is a guarantee that British aggression can not be made effectual for the ruin of American establishments. Assuredly these are valid and cogent reasons why our manufacturers want Protection, and why it should be both granted and maintained. More than that, these are equally strong reasons why American consumers should sustain and continue the beneficent tariff system now in operation.

CHAPTER XXIV.

FARMERS' SUPPLIES CHEAPENED BY PROTECTION.

"BUT for the high import duties imposed by the tariff, farmers' supplies would be as cheap as before the war." This is a constant assertion by the Free Trade press—an assertion based upon imaginary facts. The truth is all the other way. Never before in our history has agricultural produce or its money proceeds had so much purchasing power for the finished products of the loom, the anvil, or the shop.

One of the largest reaper manufacturers in the United States gives us the following facts in regard to reapers. During the four years 1857–60 his establishment made a single bar and single wheel reaper which was sold for $155. The commission for selling was $10 on each machine. Much of the labor was paid for at 75 cents a day. In 1872 he sold a double bar and double wheel reaper for $200, in 1873 for $200, in 1874 for $195, in 1875 for $185, and paid $30 dollars for selling. He now pays $1.50 for the same class of labor that was obtained for only 75 cents before the war. The reaper of the present day would cost considerably more to manufacture, if material were the same price as in 1857, because it consists of more pieces, is built more strongly, is more thoroughly finished, works more easily, and does better service than the reapers made before the war, and is estimated by the maker to be worth at least 50 per cent. more.

The following statement relative to plows and cultivators appeared in the Chicago *Evening Journal*, Nov. 5.

In a recent interview with the manager of the largest implement manufactory in the Northwest (Messrs. Furst & Bradley), he said: "It is impossible to compare the price of plows in1857–58–59–60 with the price during the past four years. There is no more resemblance between a plow of those years and now than there is between daylight and darkness. A plow, such as we sell now, would then have cost 100 per cent. more to manufacture than it costs now, with such facilities as we then had. As a general rule, the prices of agricultural implements are nearly the same as before the war, but the goods now sold are worth 100 per cent. more. They are of better quality, more convenient, will last longer, are better adapted to the wants of the farm, and are in every respect cheaper than the goods sold before the war; but a mere comparison of our price catalogues for those years with those of the past four years would be unfair, for they indicate no change in quality. We have been buying stock at a continual decline in price for several years past, but the improvements which have been made each year more than offset this decline. We are continually putting more work and better work upon our implements."

On going out we met one of the partners of the concern. He confirmed the above statement of facts, and added: "I make the knock-down argument that we will take such farm produce as we consume at the prices which prevailed before the war, and sell goods at ante-war prices. Any farmer who wants implements for his own use can come here any day and get them on these terms." When the improved quality of the implements is considered, this offer, by a responsible party, is effectually conclusive as to the relative value of farm produce and agricultural implements.

The following list of prices has been compiled, for 1875, from the prices current of the day, and for 1857 from the books of old mercantile firms, the quotations in each case being at retail in the West:

	1857.	1875.
Calicoes, prints, per yard	.12½	.08
Delaines, per yard	.25	.15
Alpacas, common, per yard	.75	.35
Alpacas, finer quality, per yard	$1.00	.40
Alpacas, finest quality, per yard	1.25	.45
Woolen cloth (shoddy in 1857), good now, per yard	1.00	.60
Woolen cloth, better quality (shoddy in 1857), good now, per yard,	1.25	.75
Woolen cloth, best quality, per yard	1.50	$1.20
Shawls, each	12.00	8.00
Cotton print cloth, per yard	.08	.04½
Raw Cotton, per pound	.13	.15
Spool silk, 100 yards, per dozen	1.25	1.25
Skein silk, per ounce	7.50	7.00
Knit undershirts, each	1.25	.50
Knit drawers, each	1.25	.50

Here, as nearly as practicable, the comparison is between the same classes and grades of goods, showing clearly that prices are less under Protection than they were under partial Free Trade. Nearly all kinds of flannels will make a similar showing. Furniture is cheaper in 1875 than it was in 1857 for like goods, or a much better article can be had for the same money. Let farmers appeal to their own memory for confirmation of what we say.

We give another table, comprising a list of *all* the woolen fabrics dealt in by a great commission house—a list which will show that, even with labor paid higher wages, with taxes much higher, and with wool protected, woolens were cheaper in 1870 than they were in 1860, making allowance for the difference between paper and gold. This table was furnished at the request of the New York *Tribune*, some years ago, by S. W. Fay, of Perry, Wendell, Fay & Co., commission merchants in New York, Boston and Philadelphia. Since then there has been a progressive decline in the prices of these goods:

STANDARD WOOLENS.	Prices in gold, 1860.	Prices in currency, 1870.
Ravine mixed, all-wool doeskins, 12 to 13 ounces	.80 to .85	.80 to .85
Munson satinets (standard article)	.50 to .55	.50 to .55
Staffordville satinets	.60 to .65	.60 to .65
Printed satinets	.25 to .62½	.25 to .62½
10-4 Holland blankets, 4¼ lbs	$3.25	$5.00
11-4 Holland blankets, 4½ lbs	4.75	6.75
"Talbot" R. scarlet flannel, ¾ (standard)	.26	.30
"F. & C." twill scarlet flannel, ¾ (standard)	.34	.45
Richmond Kentucky jeans (standard)	.26	.32½
Washington Kentucky jeans (standard)	.27	.34
Palestro Kentucky jeans	.14	.14½
St. Lawrence plaids	.21 to .22	.23½
Leicester tweeds (standard)	.62½	.65
Waterloo blanket shawls	8.00	7.50
Belvidere 9-4 printed table covers	.87½	.92½
Mixed and plain cashmerettes	.50 to .55	.52½ to .60
Boys' all-wool checks	.55 to .60	.57 to .62
Shaw diagonal all-wool cassimeres, 10 to 11 ounces	.80 to .85	.75 to .80
Evans double and twist, 8 ounces, all wool	.75	.80
Middlesex sackings	1.10	1.25
Middlesex doeskins	1.05	1.15
Middlesex shawls	7.00	7.00
Washington sackings	1.05	1.15
Glenham sackings	1.05	1.10
Glenham repellants	1.15	1.20

These figures falsify the assertion made by the Chicago *Tribune*, that, on account of our Protective policy, "we pay an average tax of 46 per cent. on everything we eat, drink, wear, or otherwise consume." It is thus evident that the great mass of the American people, particularly Western farmers, who, despite the depression of general industry caused by the panic of 1873, obtain higher prices than in 1860 for their produce, have, as consumers, no reason to complain of the existing prices for woolen goods.

The following extracts from a letter written from Iron Mountain, Mo., to the editor of the St. Louis *Tribune*, July 6, 1870, by an old man nearly 70, contain some facts well worth recollecting, and which require no comment.

I remember that fifty years ago farmers could get only 25 cents per pound for their wool, but had to pay $8 or $10 per yard for their broadcloths. That was because we had but few woolen factories in this country. What is the fact now?

We have hundreds of factories, and the consequence is that farmers are getting 50 cents a pound for their wool, and paying only $2 to $5 for their broadcloths. And yet these unscrupulous, false sheets tell them that they are obliged to sell their products for half of their value, and pay twice as much as they ought for their manufactured goods, when the very reverse of this is true, and they are actually getting twice as much for their products and paying only half as much for manufactured articles as they did fifty years ago, when we were dependent on England.

I remember that fifty years ago, before we had a cotton mill in this country, common calicoes sold for twenty-five to forty cents per yard, and a young woman at domestic service would get but seventy-five cents per week for her wages, and it would take a month's earnings to pay for a common calico dress. But what is the fact now, since we have got cotton mills of our own? Why, the price of calico is down to ten to fifteen cents a yard, and any young woman with a month's earnings can buy half a dozen calico dresses. In this connection let me state another fact for the attentive consideration of our farmers. If it had not been for the American Protective tariff we should not have had a cotton mill, nor a woolen factory, in this country to this day. We are indebted to that system for the innumerable manufacturing establishments that have sprung up in the Eastern States and made them so wealthy, and which are now springing up as if by magic throughout the West. But for that system we should have been the slaves of England still, compelled to pay her former higher prices for all manufactured goods, and to receive the former low prices, or whatever she chose to pay, for our raw products. Our fathers wearied of the galling vassalage, and threw off the yoke; and let there be none so mean and servile as to seek to subject their necks to that galling yoke again.

I remember when farmers used to get 40 cents per bushel for wheat, and 12½ cents per bushel for corn and potatoes, and 6 cents per dozen for eggs, and for fat, well-dressed chickens and other things in proportion. But now, just think of the altered state of things, down here in Southeast Missouri, all along the line of Mr. Allen's Iron Mountain road, where the iron furnaces and foundries have sprung up. Our farmers are unable to supply the market with their products, and are getting $1 a bushel for all the corn they can raise, $1.10 for potatoes, 30 to 40 cents for butter, and everything else in proportion. Should not these villainous sheets be ashamed to utter such falsehoods? Do they think they mislead our farmers? They might as well attempt to change the course of water, or dam up the Mississippi and make it run up stream.

www.ingramcontent.com/pod-product-compliance
Lightning Source LLC
LaVergne TN
LVHW021427110826
845150LV00007B/2114

* 9 7 8 1 4 2 5 5 0 7 4 6 6 *